Patrick Moore's
Practical Astronomy Series

Springer
London
Berlin
Heidelberg
New York
Hong Kong
Milan
Paris
Tokyo

Other titles in this series

The Art and Science of CCD Astronomy
David Ratledge (Ed.)

The Observer's Year
Patrick Moore

Seeing Stars
Chris Kitchin and Robert W. Forrest

Photo-guide to the Constellations
Chris Kitchin

Software and Data for Practical Astronomers
David Ratledge

The Sun in Eclipse
Michael Maunder and Patrick Moore

Software and Data for Practical Astronomers
David Ratledge

Amateur Telescope Making
Stephen F. Tonkin

Observing Meteors, Comets, Supernovae and other Transient Phenomena
Neil Bone

Astronomical Equipment for Amateurs
Martin Mobberley

Transit: When Planets Cross the Sun
Michael Maunder and Patrick Moore

Practical Astrophotography
Jeffrey R. Charles

Observing the Moon
Peter T. Wlasuk

Deep-Sky Observing
Steven R. Coe

AstroFAQs
Stephen F. Tonkin

The Deep-Sky Observer's Year
Grant Privett and Paul Parsons

Field Guide to the Deep Sky Objects
Mike Inglis

Choosing and Using a Schmidt–Cassegrain Telescope
Rod Mollise

Astronomy with Small Telescopes
Stephen F. Tonkin (Ed.)

Solar Observing Techniques
Chris Kitchin

Observing the Planets
Peter T. Wlasuk

Light Pollution
Bob Mizon

Using the Meade ETX
Mike Weasner

Practical Amateur Spectroscopy
Stephen F. Tonkin (Ed.)

More Small Astronomical Observatories
Patrick Moore (Ed.)

Observer's Guide to Stellar Evolution
Mike Inglis

How to Observe the Sun Safely
Lee Macdonald

Astronomer's Eyepiece Companion
Jess K. Gilmour

Observing Comets
Nick James and Gerald North

Observing Variable Stars
Gerry A. Good

Visual Astronomy in the Suburbs
Antony Cooke

Telescopes and Techniques
An Introduction to Practical Astronomy

Second Edition

Chris Kitchin

With 127 Figures

Springer

Cover illustration: The Orion Nebula, M42. (Michael Stecker/Galaxy Picture Library)

British Library Cataloguing in Publication Data
   Kitchin, Christopher R. (Christopher Robert), 1947–
   Telescopes and techniques: an introduction to practical
   astronomy. – 2nd ed. – (Patrick Moore's practical astronomy
   series)
   1. Telescopes 2. Astronomy
   I. Title
   522.2
ISBN 1852337257

Library of Congress Cataloging-in-Publication Data
Kitchin, C. R. (Christopher R.)
   Telescopes and techniques: an introduction to practical
   astronomy/C.R. Kitchin.
      p. cm. – (Patrick Moore's practical astronomy series)
   Includes bibliographical references.
   ISBN 1-85233-725-7 (alk. paper)
   1. Astronomy. I. Title. II. Series.
QB43.3.K58 2003
522–dc21   2003045429

Patrick Moore's Practical Astronomy Series ISSN 1617–7185
ISBN 1–85233–725–7 2nd edition Springer-Verlag London Berlin
   Heidelberg
ISBN 3–540–19898–9 1st edition Springer-Verlag Berlin Heidelberg
   New York
ISBN 0–387–19898–9 1st edition Springer-Verlag New York Berlin
   Heidelberg
a member of BertelsmannSpringer Science+Business Media GmbH
http://www.springer.co.uk

First published 1995
Second edition 2003

Typeset by EXPO Holdings, Malaysia
58/3830-54321   Printed on acid-free paper   SPIN 11018780

For
Spruce, Misty, Ryan, Benji, Merlin and Bassett
and in memory of
Pip, Badger, Wills, Jess, Chalky, Midnight, Sheba, TC, Satchmo, Monty and
Snuffles

Preface to the First Edition

This book arose from the need to introduce first-year astronomy students at the University of Hertfordshire to the basic techniques involved in using a telescope and finding their way around the sky. It soon transpired that many aspiring astronomers, not just those in a university environment, needed similar guidance. The aim of the book is therefore to be introductory in the sense that prior knowledge is not assumed, but not in the usual sense that mathematics is avoided. Anyone wanting to learn about telescopes, how they work, how to use them, and how to choose a telescope for their own use, should find this book helpful. It provides the information on how to set up a telescope from scratch, to find objects in the sky, both those bright enough to be seen with the naked eye, and those for which the telescope must be set on to the object before it can be seen. It explains such things as sidereal and solar time, right ascension and declination, light grasp, aberrations, etc. in sufficient detail for useful work to be undertaken. The techniques involved in visual work with a telescope and imaging with photography and CCDs are also explained to a similar depth. Ancillary work and instrumentation such as data processing, photometry and spectroscopy are outlined in somewhat less detail. I hope readers find the book useful and as interesting as I found it to write.

C.R. Kitchin
1995

Preface to the Second Edition

In the seven years since the first edition was produced, *Telescopes and Techniques* has shown itself to fill a requirement for an explanation of all aspects of the use of telescopes and detectors, etc. However, it has also become apparent that some topics that were omitted first time around, such as radio telescopes, should have been included. Many techniques, especially relating to detectors, have also undergone rapid changes over that interval.

This second edition has therefore been expanded to include a good deal of new material, and up-dated. I hope that it will continue to be useful to astronomers of all kinds, and to provide the background to operating telescopes successfully.

May you all have clear skies!

C.R. Kitchin
2003

Contents

Section 1: **Telescopes**

1 Types of Telescope 3
   Historical Introduction 3
   Modern Instruments 16
      Mounting 17
      Schmidt Camera 18
      Multi-Mirror and Space Telescopes 21
      Atmospheric Compensation 23
   Radio Telescopes 25
      Interferometers 27

2 Telescope Optics 31
   Point Sources 31
   Extended Images 35
   Objectives 38
   Eyepieces 41
      Accessories 45
   Aberrations 48
   Interferometers 52
   Mountings 54
   Observatories and Observing Sites 58
   Exercises 59

3 Modern Small Telescope Design 61
   Introduction 61
   Making Your Own 63
   Commercially Produced Telescopes 67
   Binoculars 68

Section 2: **Positions and Motions**

4 Positions in the Sky 73
   Spherical Polar Coordinates 73
   Celestial Sphere 75
   Altitude and Azimuth 80

Rotation. 81
Solar and Sidereal Days 82
Declination and Hour Angle. 82
Time 85
　　Mean Solar Time. 85
　　Solar Time 86
　　Civil Time 87
　　Sidereal Time. 87
Right Ascension and Declination.. 90
Other Coordinate Systems. 92
Heliocentric Time 93
Julian Date 93
Spherical Trigonometry 95
Exercises 99

5　**Movements of Objects in the Sky** 101
Diurnal Motion 101
　　Circumpolar Objects 102
Seasons and Annual Motions. 103
Movement of the Moon and Planets. 106
　　Moon 107
　　Other Solar System Objects 108
Proper Motion. 111
Precession 111
Parallax. 113
Aberration 114
Relative Planetary Positions 116
　　Position with Respect to the Earth 116
　　Eclipses. 117
　　Position in an Orbit. 121
　　Synodic Period 121
Exercises 123

6　**Telescope Mountings** 125
Introduction. 125
Equatorial Mountings. 126
Alt-Az Mountings 129
Making Your Own Mounting 129
Alignment 130
Setting Circles 132
Guiding. 133
Modern Commercial Mountings. 135

Section 3: **Observing**

7　**Electromagnetic Radiation**. 139
Introduction. 139

Intensity 140
Photons 141
Polarisation 141
Range. 142
Measurements. 143
Photometry 144
Spectroscopy 144
Polarimetry 144

8 **Visual Observing** 145
Introduction. 145
General and Practical Considerations and
Safety. 146
Finding 147
Moon. 150
Planets.. 154
Sun 158
Finding 158
Observing 159
Stars. 162
Stellar Nomenclature. 162
Magnitudes 165
Observing Stars. 167
Nebulae and Galaxies. 171
Daytime Observing 173
False Observations 174
Exercise 175

9 **Detectors and Imaging** 177
The Eye. 177
Charge Coupled Devices (CCDs). 180
Image Processing. 184
Photography. 186
p-i-n Photodiode 189
Superconducting Tunnel Junction Detectors. . 190
Exercise. 190

10 **Data Processing**. 193
Introduction. 193
Data Reduction 194
Data Analysis. 197
Linear Regression 198
Correlation Coefficient. 200
Student's *t* Test.. 203
Exercises 205

11 **Photometry**. 207
Introduction. 207

CCD Photometry. 207
Photographic Photometry 208
Absolute Magnitude 210
Wavelength Dependence. 211
UBV System. 212
Bolometric Magnitude. 213
Spatial Information. 214
Photometers. 214
Observing Techniques 215
Exercises. 216

12 **Spectroscopy** 219
Introduction. 219
Spectroscopes.. 220
Spectroscopy 222
Spectral Type 222
Luminosity Class 225
Radial Velocity 225
Spectrophotometry. 226
Exercise. 227

Appendix 1: **Telescope, Detector and Accessory
Manufacturers and Suppliers.. . . . . 229**

Appendix 2: **Bibliography** 233

Appendix 3: **National and Major Astronomical
Societies**. 237

Appendix 4: **Constellations**. 243

Appendix 5: **Answers to Exercises.** 245

Appendix 6: **SI and Other Units** 249

Appendix 7: **Greek Alphabet.** 251

Index. 253

Section 1

Telescopes

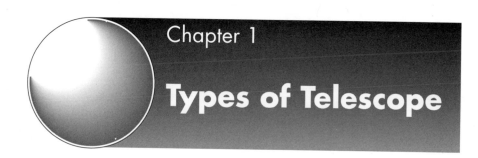

Chapter 1

Types of Telescope

Historical Introduction

Until recently, the invention of the telescope was generally attributed to a Dutch spectacle maker called Hans Lippershey (or Lippersheim, 1570?–1619). He worked at Middleburg on the island of Walcheren, some 60 km north-west of Antwerp. The probably apocryphal story has it that in 1608 his children discovered, while playing with some of his spare lenses, that one combination made a distant church spire appear much closer. The exact combination of lenses they and he used is no longer known, but it is likely to have been a pair of converging lenses. An example of the resulting telescope was duly presented to the States-General, Prince Maurice. The news of the discovery spread rapidly, reaching Venice and Galileo only a year later. The details received by Galileo just concerned the effect of observing with the instrument, and not the details of its design. However, he had at that time been working extensively on optics, and in a few hours was able to design an optical system that reproduced the reported distance-shortening effect of the instrument. Galileo's design used a long focal length converging lens and a shorter focal length diverging lens placed before the focus (Figure 1.1), and had the advantage for terrestrial use of producing an upright image. This optical system we now know as a Galilean refractor after its inventor, and it still finds use today in the form of opera glasses.

The invention of the telescope, however, may well predate Lippershey. Lately, historical research has sug-

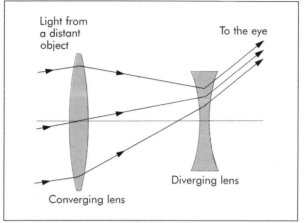

Figure 1.1. Optics of the Galilean refractor.

Within the figure:
Light from a distant object
To the eye
Diverging lens
Converging lens

gested that a form of telescope may have been discovered by the Englishman Leonard Digges (?–1571?), who also invented the theodolite, some time around 1550. His design appears to have used a long focal length lens and a short focal length concave mirror for what we would now call the eyepiece, and therefore was probably used to look at objects behind the observer (Figure 1.2). It is possible that a garbled account of such a system could have led to the idea of the crystal ball used by fortune-tellers. Even earlier telescopes may be ascribable to Roger Bacon (1220–92, who cer-

Figure 1.2.
Suggested optical system of Digges' possible telescope.

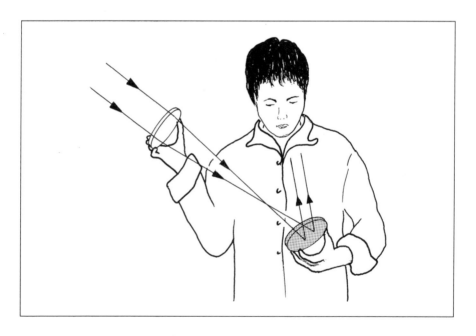

tainly knew about spectacles for the correction of long-sight), and to Giambattista della Porta (1535–1615, who investigated the *camera obscura)*, but these attributions are much more dubious.

Galileo's first telescope magnified by about three times, but later he produced instruments with magnifications up to 30. Pointing his telescopes towards objects in the sky revealed the craters on the Moon, that planets, unlike stars were not point sources, and that the four largest satellites of Jupiter, which he called the Medicean stars, but which we now know as the Galilean satellites, orbited that planet. Most importantly, he observed the phases of Venus, and its angular size changes to prove that it at least orbited the Sun and not the Earth, so giving strong support to the heliocentric theory of Copernicus. However, the magnification of the Galilean refractor was limited in the early seventeenth century by the difficulty of producing short focal length diverging lenses with their deep concave surfaces, and so was soon replaced for astronomical purposes by the astronomical refractor. This design is still in use today and has a long focal length converging first lens (or objective), followed by a much shorter focal length converging lens as the eyepiece, placed after the focus of the objective (Figure 1.3). It gives an upside-down image of course, so limiting its use for terrestrial purposes, but causing little disadvantage when looking at objects in the sky.

Although the astronomical refractor could more easily be made to give higher magnifications than the Galilean telescope, both designs suffered from other problems, which considerably limited their use. These problems are now known as *aberrations,* and are discussed in more detail in Chapter 2. Two of the six primary aberrations affected these early telescopes particularly badly, and

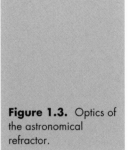

Figure 1.3. Optics of the astronomical refractor.

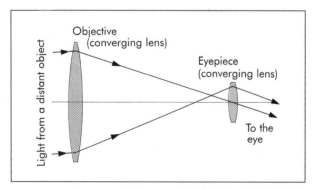

Objective (converging lens)

Eyepiece (converging lens)

Light from a distant object

To the eye

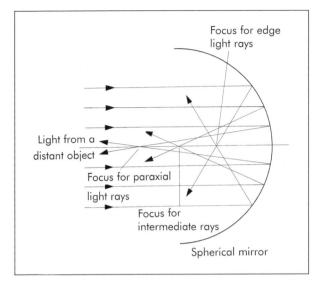

Figure 1.4.
Spherical aberration in a mirror: rays from different radii of the mirror come to foci at different points along the optical axis.

these were spherical aberration and chromatic aberration. Spherical aberration derives its name from the way in which a spherical mirror brings parallel light rays to a "focus" (Figure 1.4). The light rays at different distances from the optical axis are brought to foci at different distances along the optical axis. Simple lenses produce a similar effect (Figure 1.5). Thus there is no point at which the eyepiece can be placed to give a sharp image.

Chromatic aberration is a similar effect in which light rays of differing wavelengths are brought to differing foci along the optical axis (Figure 1.6). Since the law of reflection is independent of wavelength,

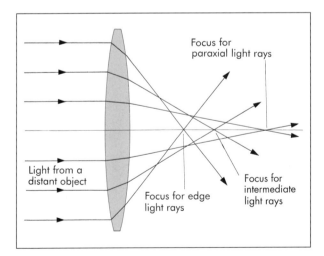

Figure 1.5.
Spherical aberration in a simple lens.

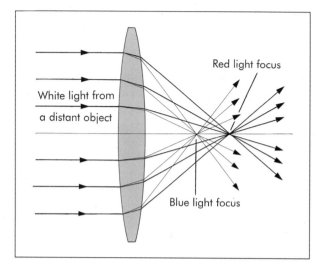

Red light focus

White light from

a distant object

Blue light focus

Figure 1.6.
Chromatic aberration
in a simple lens: for
simplicity the effects of
spherical aberration
have been omitted.

however, this aberration applies only to lenses. Like spherical aberration, there is no point at which the image will be sharply in focus. However, since the eye has a sensitivity that peaks in the yellow-green part of the spectrum, on focusing a telescope in which chromatic aberration is present, the tendency is to put the yellow-green part of the image into the sharpest focus.

In addition to these two aberrations, the quality of glass in the seventeenth century was very poor. Lenses could thus contain debris from the furnace in which the glass had been melted, bubbles of air, unannealed stresses, etc. The surfaces of the lenses could also be poorly polished and deviate markedly from the correct spherical shape. The images in these early telescopes were thus of very poor quality indeed by today's standards. Saturn, for example, was not observed correctly as a planet surrounded by rings until Christian Huyghens' (1629–95) work in 1656, almost 40 years after the application of the telescope to astronomy. Prior to Huyghens, Saturn had been variously observed as a triple planet, an elliptical planet and a planet with handles. If we add to the poor quality of the images the difficulty of finding and tracking objects owing to the tiny fields of view of the telescopes and their inadequate or non-existent mountings, then the discoveries made by the early telescopists become truly remarkable. It also becomes much less surprising that Galileo's discoveries and their support for the Copernican model of the solar system were not accepted immediately and universally. Sceptics looking through one of his instruments could easily, and with some justification, come to quite different interpretations of the faint fleeting blurry images from those reached by heliocentrists.

The problems caused by chromatic and spherical aberrations were soon found to be reduced if the focal length of the lens was long compared with its diameter, or, as we would now express it, if lenses with high focal ratios were used. This led to increasing lengths for telescopes, and to the use of nested draw-tubes so that the instrument could be collapsed to a convenient length when not in use. Even for terrestrial telescopes (which were provided with an upright image by the addition of a relay lens, Figure 1.7), lengths of 3–5 metres became common. For astronomical work, this trend eventually, in the late seventeenth century, resulted in the aerial telescope (aerial in the sense of "belonging to the air", not today's common meaning of a radio antenna). Johannes Hevelius of Gdansk (1611–87), Huyghens, and others constructed telescopes up to 60 metres long. Telescopes of such a size could not take the same form, nor be handled in the same way, as smaller instruments. In one version, the objective would be mounted in a small tube near the top of a mast. The eyepiece would be in a separate tube, and the two tubes connected and aligned by a taut line. The eyepiece was held in the hand, and the observer moved around the mast to keep the line in tension, and to look at different parts of the sky. When required, the objective could be raised or lowered on the mast by assistants to make a major change in the altitude to be observed. It was the open nature of these telescopes that led to their name. Such instruments would clearly be incredibly difficult

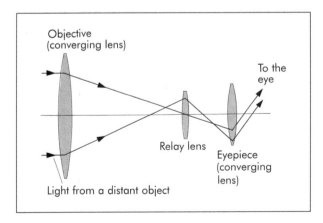

Figure 1.7. Simple terrestrial telescope (upright image).

to use, swinging and trembling in the lightest of breezes, and with very tiny fields of view. Nonetheless they could be used to great effect; Giovanni Cassini (1625–1712), for example, found four new satellites of Saturn, and the break in the rings of Saturn still known as Cassini's division.

Since reflection, unlike refraction, is independent of wavelength, an alternative approach to overcoming the problem of chromatic aberration using mirrors in place of the main lenses was suggested in the mid-seventeenth century. It was also realised that by deepening the curve of the primary mirror from a spherical to a paraboloidal shape, spherical aberration could additionally be eliminated (Figure 1.8). James Gregory (1638–75) of Aberdeen proposed the first such design in 1663. His system, now called the Gregorian telescope, used a paraboloidal mirror as the objective (usually now called the primary mirror), and a small ellipsoidal mirror placed after the focus of the primary mirror (the secondary mirror). Light from a distant object hitting the primary mirror would be reflected to the secondary mirror that would reflect it down the telescope tube again to a second focus. The light would emerge from the telescope through a small central hole in the primary mirror and be observed with an eyepiece made with lenses (Figure 1.9). Gregory had a telescope made to his design by two London-based opticians, but was dissatisfied with the results, and there is no record of any successful observations ever being made with the instrument.

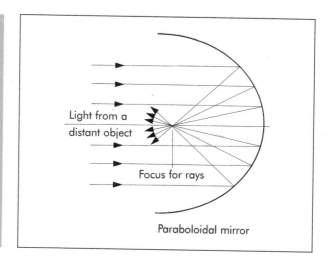

Light from a
distant object

Focus for rays

Paraboloidal mirror

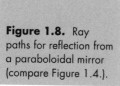

Figure 1.8. Ray paths for reflection from a paraboloidal mirror (compare Figure 1.4.).

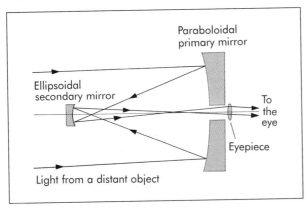

Figure 1.9.
Gregorian telescope.

The first working reflector was therefore designed and built by Isaac Newton (1642–1727) five years after Gregory's work. Newton's telescope was of a different design from the Gregorian, with a parabolic primary but a flat mirror for the secondary set at 45° to the axis of the telescope. The light was thus brought to a focus at the side of the telescope and that is where the eyepiece had to be positioned (Figure 1.10). Newton's system is still in widespread use today and called the Newtonian telescope. It is mostly found in the smaller telescopes by today's standards, these being used and often made by amateur astronomers. The first Newtonian telescope, however, had a diameter of only just over an inch (25 mm), so even the smallest amateur's telescope today dwarfs the one that Newton used. All the merits of Newton's design were not imme-

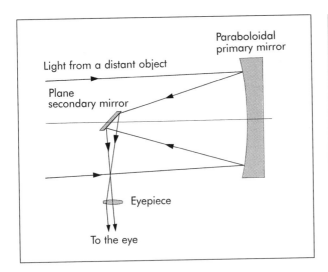

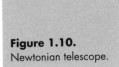

Figure 1.10.
Newtonian telescope.

diately obvious in practice, because the mirrors he used were spherical, not parabolic, and made from speculum metal (an alloy of variable composition but normally about 75% copper, 25% tin, plus sometimes some zinc and a little arsenic, also known as bell metal). Speculum metal has a reflectivity of only about 60% when freshly polished, and much less when tarnished, which happens quite quickly. Thus with an aperture of just over an inch, and two such mirrors, Newton's telescope would only have delivered to the eye a little more light than could be obtained looking directly at the object. Since the instrument had a magnification of about 25 times, then except for point sources like stars, this light would be spread over a 600 times greater area, making objects seen through the telescope appear very dull and faint, and so only the brightest objects could have been observed.

In 1672, only four years after Newton's invention, yet another design for a reflecting telescope was produced, this time by the Frenchman Guillaume Cassegrain, about whom little is known. This optical design, or variations on it, is still almost universally used for large telescopes. Both the Hubble Space Telescope and the Keck 10 m telescope are of this basic design. It is very similar to the Gregorian telescope, but with the elliptical concave secondary replaced by a convex hyperboloidal mirror, and with that mirror placed before instead of after the focal point of the primary mirror (Figure 1.11). A major reason for the continuing popularity of the Cassegrain design is its telephoto property. That is, it has a focal length much greater than its physical length. This arises, as can be seen from Figure 1.11, because the secondary mirror produces a much narrower converging cone of light than that originally

Figure 1.11.
Cassegrain telescope.

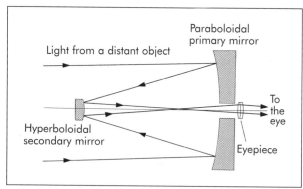

coming from the primary. The short length that this gives to the telescope reduces the cost of the instrument by far more than might be expected. There is an economic gain at every point in the overall design; the shorter tube length means it can be made from thinner materials, the mounting and drive can be smaller and lighter, the dome can be smaller, and so on. The Gregorian design has a similar telephoto effect, but with less of a reduction in overall telescope length because the secondary mirror is placed after the focus of the primary, not before it. The Gregorian also suffers through having a very small field of view, making it difficult to use in comparison with other designs. Nonetheless, in the eighteenth century, it was the Gregorian telescope which was the more popular of the two designs, and instrument makers such as James Short (1710–68) and others made numerous examples with apertures of a few inches. The popularity of the design at that time arose from its relative ease of manufacture and testing. The concave secondary of the Gregorian was much more straightforward to produce than the convex secondary of the Cassegrain. The first reflectors of any design to be made with correctly shaped paraboloidal primaries were produced by John Hadley (1682–1744) in the early 1720s.

Much later, William Herschel (1738–1822), in his large telescopes, dispensed altogether with the secondary mirror in order to eliminate its light losses. He simply tipped the primary mirror to one side and placed the eyepiece immediately after the primary focus. This type of design has become known as a Herschelian telescope. In the form that Herschel used, the images it produces are seriously degraded by aberrations. Not only does spherical aberration reappear, but also other aberrations such as coma and astigmatism (Chapter 2) become significant. Herschel's problems with the design arose because at that time it was only possible to produce symmetrically shaped mirrors. Today the design is in widespread and successful use, not least at microwave wavelengths for the ubiquitous satellite television receiving aerials, but with asymmetrical, or off-axis mirrors.

In the early eighteenth century, however, small reflectors and small refractors were of roughly equal (and fairly poor) quality. In 1729 Chester Moor Hall (1703–71) invented the achromatic lens, but it was not until 1754 that the situation for refractors was transformed when John Dollond (1706–61) began manufacturing high quality achromatic lenses commercially. This invention reduced the effects of chromatic aberration on images by a large amount. It relied for its effect on the fact that the light deviating property of a glass (refraction) and the light splitting property of a glass (dispersion) vary from one type of glass to another, but not always by equal amounts. Thus it was possible to find two varieties of glass that were such that if one were made into a converging lens, and the second into a diverging lens, then when they were combined together the dispersion of one was cancelled out by the other, but the overall combination still acted as a converging lens. The glasses chosen by Dollond remain in common use today, and were crown glass for the converging lens, and flint glass for the diverging lens. The combination of the two lenses is called an *achromatic doublet*. One of the most widespread designs for such a doublet has the two inner (contact) surfaces with the same radius so that the two lenses may be cemented together to produce a robust combination not easily misaligned (Figure 1.12). However, many variations on this basic theme are possible, including separating the two components by a small distance, and with care some of the other aberrations can also be reduced. A two-element achromat typically reduces the chromatic aberration by a factor of ten, but does not eliminate it completely. The coloured images of the simple lens are folded around so that at any point on the optical axis two wavelengths are simultaneously in focus (Figure 1.13). Choosing the coincident wavelengths appropriately can produce a lens optimised for a particular purpose. Thus for visual work, wavelengths around 400 and 600 nm should coincide, while for photographic work, having 350 and 500 nm together may be more useful. A third lens may be added to the combination, and this folds the colour spread a second time, reducing the remnant of chromatic aberration even further, and producing a lens called an *apochromat*. Yet more lenses may be added to continue to improve the colour and other aberrations in the image, but these are not economically viable for the large lenses required for telescope objectives; modern camera and other lenses, however, may go on to have a dozen or more components.

With the invention of the achromat for the objective, and the use of multi-lensed eyepieces (see later), we have the refractor in essentially its modern form

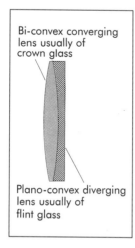

Bi-convex converging lens usually of crown glass

Plano-convex diverging lens usually of flint glass

Figure 1.12.
Achromatic doublet.

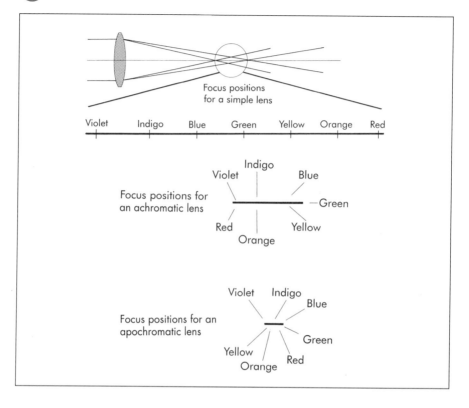

(Figure 1.14). From the mid-eighteenth century onwards, the refractor was developed rapidly. The other major contribution to its advance came from the German astronomer, Joseph von Fraunhofer (1787–1826) who developed greatly improved glass making techniques, thus providing large, high-quality blanks for the production of the lenses. The development of the refractor culminated at the end of the nineteenth century in the 1 m refractor at Yerkes. The refractor at this size is, however, limited by several factors: the lenses are several inches thick, thus becoming very heavy and absorbing light to a significant extent; they can only be supported at their edges and can sag under their own weight; the residual chromatic aberration becomes intrusive; and, by no means finally, the long focal ratios required to reduce other aberrations produce very lengthy telescopes requiring massive mountings and huge and expensive domes.

While refractors were being pushed towards their limits, reflectors also continued to be used, because

Figure 1.13.
Chromatic aberration for a simple lens, an achromatic doublet and an apochromat.

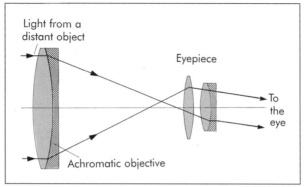

Figure 1.14. Modern astronomical refractor.

mirrors could be made far larger than lenses. This was due to several factors: firstly the casting of speculum metal, though by no means trivial, was easier than that of glass; secondly, even if the casting did contain bubbles, etc. this was of relatively little importance compared with the glass for a lens, because light does not pass into the material forming a mirror, so only its surface needs to be of high quality; finally, mirrors can be supported on their backs, as well as at the sides, so distortion due to gravitational loading is less significant. Thus even with the poor reflectivity of speculum metal, the total light gathered could be superior to that of a refractor. William Herschel, for example, produced his 1.2 m reflector in 1787, over one hundred years before the rather smaller Yerkes refractor, and the largest ever speculum-mirror based telescope, a 1.8 m Newtonian reflector, was built by William Parsons (1800–67, third Earl of Rosse) in 1845.

Speculum metal mirrors suffer from an additional serious drawback apart from their low reflectivity – repolishing to remove the tarnish will also spoil the precise shape of the surface of the mirror. When mirrors were of poor optical quality anyway, this was not of great importance; but once their surfaces could be shaped (figured) closely to the required paraboloid, then repolishing became a very lengthy process – perhaps half as long as producing the original mirror from its blank in the first place. Thus William Parson's telescope (which made some significant discoveries, including the spiral nature of some galaxies) was the culmination of its type.

Modern Instruments

By the latter part of the nineteenth century, metal-on-glass mirrors replaced the speculum metal used for the previous two centuries. These had the advantages of high reflectivities (95% or more), and the fact that renewing a tarnished surface could be done by removing the old reflecting layer with a chemical solution, leaving the glass and its precisely shaped optical surface untouched, and then adding a new coating of the metal. Initially, silver was used as the reflector, and this could be deposited chemically. However, silver tarnishes very quickly, necessitating new coatings at intervals of a few weeks, so that most telescopes now use a layer of aluminium evaporated on to the glass surface while it is inside a vacuum chamber. Aluminium oxidises, but the thin layer of aluminium oxide produced when the mirror is first exposed to the atmosphere is transparent, and furthermore seals the surface, protecting the aluminium from additional oxidation. Nowadays, it is common practice to overcoat the aluminium with a further protective layer of silicon dioxide. Even so, most major telescopes have to have their mirrors recoated at intervals of a year or so.

By the early part of the twentieth century, most new major telescopes being built were recognisably the ancestors of today's instruments – that is, they were metal-on-glass reflectors of the Cassegrain design or one of its derivatives such as the Ritchey–Chrétien. In the latter, the primary mirror is deepened from a paraboloid to a hyperboloid shape, while the secondary remains a hyperboloid, but with a steeper curve. This change slightly degrades the on-axis image, but not outside the diffraction limits of the telescope (Chapter 2), but very considerably improves the image away from the optical axis. Thus telescopes of the Ritchey–Chrétien design can produce good quality images over a field of view several tens of minutes of arc across, compared with the basic Cassegrain design in which the field of sharp focus may be limited to only a few minutes of arc.

Most large, and many smaller telescopes can be used in several modes. One of the commonest is as a Cassegrain system, as already mentioned. Alternatively the Cassegrain secondary mirror can be replaced by a flat mirror set at 45° to produce a Newtonian. In the larger telescopes, the instrumentation to be used, such as a camera, photometer or spectroscope, may be no

Figure 1.15.
Telescope used at
prime focus.

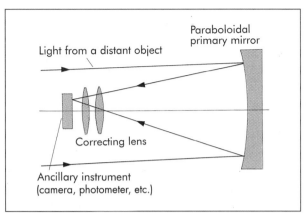

larger than the Newtonian secondary, and so that mirror may be dispensed with altogether and the instrumentation placed directly at the focus of the primary mirror. Such a system is called the primary focus of the telescope, and is similar to the Herschelian system except that it is on-axis. In telescopes over three or four metres in diameter, even the observer can occupy the prime focus position in order to operate the instrumentation without obstructing too great a proportion of the incoming light. Usually weak lenses, sometimes of complex shapes and/or made from exotic materials, have to be placed before the prime focus to produce a reasonable quality image. These are known as correcting lenses (Figure 1.15).

Mounting

All telescopes have to have a mounting to enable them to be pointed at the required part of the sky, and then to follow (track) the object being observed as it changes position because of the Earth's rotation. Little has been said about the design of mountings in the previous discussion, though they are at least as important for the *successful* use of a telescope as the optics themselves. Some mounting designs are discussed later. Here we just note that with most mountings, the position in space of the focus of the telescope changes as the telescope moves. Any instrumentation placed at that focus therefore has to be relatively lightweight, and capable of being operated at varying angles of inclination. Large ancillary instruments, or instru-

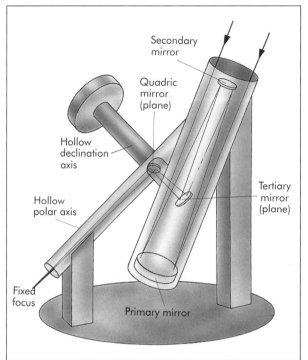

Figure 1.16. Coudé system.

ments under development, are therefore difficult to attach to telescopes. A third mode therefore exists for many major telescopes, known as the Coudé focus in the case of an equatorial mounting (Figure 1.16), or the Nasmyth focus in the case of an alt-azimuth mounting (Figure 1.17). This provides a focus that is fixed (for the Coudé system), or remains horizontal (in the case of the Nasmyth system). In both cases, therefore, large and/or delicate equipment can be used with the telescope.

Schmidt Camera

There is one exception to the almost universal use over the last century of reflectors of Cassegrain or Ritchey–Chrétien design, and that is the Schmidt camera. This was invented in 1930 by the Estonian optician, Bernhard Schmidt (1879–1935). This instrument has one great advantage over the conventional telescope – its very wide field of sharp focus. The field of sharp focus for a Cassegrain design can be as little as a few minutes of arc, that for a Ritchey–Chrétien, ten

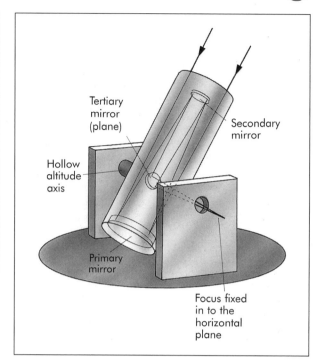

Figure 1.17.
Nasmyth system.

or twenty minutes of arc; the Schmidt camera, by contrast, can cover a field of view six to ten *degrees* across.
This wide field of view makes the instrument ideal for
survey work, since large areas of sky can be covered
quickly. The instrument does, however, have several
drawbacks: firstly, the focus is inaccessible, making it
only usable normally as a camera; secondly it has the
opposite of the Cassegrain's telephoto property in that
its physical length is at least twice its focal length; other
problems are that it uses a lens, with the attendant
difficulties of support, and that the mirror is considerably larger than the useful aperture. An instrument that
uses both lenses and mirrors in the primary light gathering capacity, like a Schmidt camera, is called a catadioptric instrument (reflectors can also be called catoptric
instruments, and refractors, dioptric instruments).

The superior optical performance of the Schmidt
camera lies in the apparently retrograde step of using a
spherical mirror in place of the parabolic mirrors of
other telescopes. This reintroduces the problem of
spherical aberration. However, the other aberrations
such as coma, astigmatism, etc. (see later) are avoided.
Spherical aberration is then eliminated by using a cor-

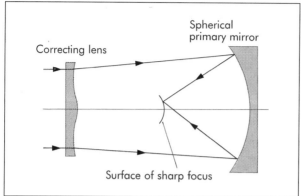

Correcting lens

Spherical primary mirror

Surface of sharp focus

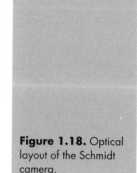

Figure 1.18. Optical layout of the Schmidt camera.

recting lens placed at the radius of curvature of the primary mirror (Figure 1.18). This lens is thin, and the chromatic aberration that it produces can be kept small enough to be negligible. The only remaining aberration therefore in the system is that the field of sharp focus is not flat (field curvature), and this is overcome by bending the photographic plate or other detector into the same shape as the sharp focus surface. The largest Schmidt camera is the 1.34 m instrument at Tautenberg in Germany. As with ordinary refractors, the size is limited by the maximum size of lens that can be given an adequate support. Although there are a number of other smaller cameras at other observatories, there is relatively little need for them since, with their wide fields of view, two major instruments (one in each hemisphere) can quickly cover the whole sky.

The original Schmidt design can only be used as a camera, However, it has been adapted to allow an accessible focus in recent years. The resulting instrument is known as a Schmidt–Cassegrain, and forms the basis for several types of excellent small telescope produced for the amateur and education markets. The Schmidt–Cassegrain uses a correcting lens close to the focus of the primary mirror, and a secondary mirror (which is often attached to the lens, so reducing diffraction effects) to send the focus out through the back of the telescope through a hole in the primary mirror, as with the Cassegrain design (Figure 1.19). A rather similar design, called the Maksutov, uses spherical surfaces for the optical components (Figure 1.20). These instruments have the telephoto advantage of the

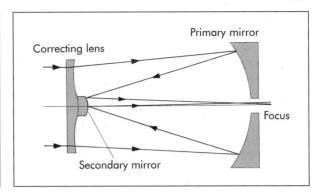

Figure 1.19. Optical layout of a Schmidt–Cassegrain telescope.

Cassegrain, allied with a wider field of sharp focus. Their production in large quantities also means that they are relatively cheap.

Multi-Mirror and Space Telescopes

The history of telescopes shows a recurring theme: existing designs are found to be limited in some way (image quality, size, cost, etc.); a new development in telescope design then occurs which supersedes the existing designs; that new design is then used as the basis of telescopes of ever-increasing size, until it too reaches its limitations in some way. We have seen this with the early refractors, then the early reflectors, refractors again, and most recently again with

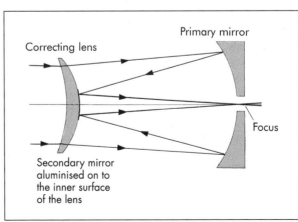

Figure 1.20. Optical layout of a Maksutov telescope.

reflectors. The current limitation is that of the difficulty of producing large mirrors. Although some recent telescopes have relatively conventional primary mirrors up to 8 m in diameter, that is nearing or at the upper limit of what is possible for current technology. The new development, which we are just seeing at its commencement, is therefore into multi-mirror telescopes. The first one of these of any significance was the Mount Hopkins instrument that used six 1.8 m primary mirrors in a relatively conventional Cassegrain design, but on a single mounting and feeding a common focus (Figure 1.21). Although this gave the equivalent of a 4.4 m telescope in area, at perhaps a third of the normal cost of such an instrument, it has not in practice proved successful, and the instrument now uses a single 6.5 metre monolithic mirror. More recently the two 10 m Keck telescopes have come into operation with much more success. These use thirty-six 2 m mirrors as the segments of a single mirror (Figure 1.22). Each segment has to be correctly positioned with respect to the others, and this is accomplished by active supports that are continually adjusted by a computer to keep the alignment within the required limits of about 50 nm. Yet another approach is to link separate telescopes by fibre optics, and this is the principle of the European Southern Observatory's (ESO) very large telescope project. This has four 8 m conventional telescopes that can operate indepen-

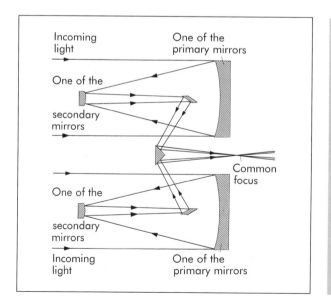

Figure 1.21. Optical layout of the Mount Hopkins multi-mirror telescope (only two of the six primary mirrors are shown).

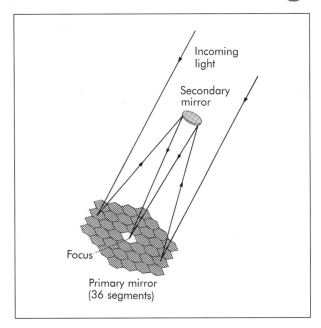

Figure 1.22. Optical layout of the Keck 10 m multi-mirror telescope.

Incoming light

Secondary mirror

Focus

Primary mirror
(36 segments)

dently of each other or be linked to give the light gathering power of a 16 m mirror and they can also operate as an interferometer (see the section on Radio Telescopes, below).

In a separate development, we have various space telescopes, such as the Hubble Space Telescope and its planned replacement, the James Webb telescope, that are designed to overcome the limitations imposed on ground-based telescopes by the Earth's atmosphere. This both degrades the image through scintillation, and absorbs radiation strongly in the ultraviolet and infrared parts of the spectrum.

Atmospheric Compensation

The problem of the atmosphere reducing the resolution of ground-based telescopes is also being addressed by active atmospheric compensation. This technique counteracts the distortion produced by the atmosphere to produce an image at the diffraction limit (Chapter 2) of the telescope.

The effect of the atmosphere is to delay some parts of the wavefront coming from the object more than others. What should be a flat wavefront therefore becomes one with bumps and hollows. A small mirror is added to the telescope whose shape is the inverse of that of the wavefront. After reflection from that mirror therefore, the wavefront is again flat. There are two problems with this procedure. One is finding the shape of the wavefront so that the correcting mirror can be made to the right profile. The second is that the distortions produced by the atmosphere change on a time scale of a few milliseconds.

The shape of the distorted wavefront is determined by monitoring the shape of the wavefront from a guide star close to the object actually being observed. There are several techniques for determining the shape of the wavefront, and one of those used most commonly is the Hartmann sensor. This consists of a grid of small lenses, each of which produces an image of the guide star on to a CCD (Chapter 9). The movements of those images from their correct positions then provide the details of the wavefront distortions. The guide star needs to be within a few seconds of arc of the object whose image is to be corrected. If such a star is not available, some compensation systems produce an artificial star by shining a powerful laser upwards to cause sodium atoms to glow at a height of about 90 km.

Once the form of the wavefront has been found, the correcting mirror is adjusted to be its inverse shape. The correcting mirror may be constructed from a number of small independent segments, or it may be very thin so that its shape can be bent. In either case the mirror or mirror segments are mounted on computer controlled actuators whose position can be changed in a fraction of a millisecond in order to produce the correct profile for the mirror.

With such an active atmospheric compensation system operating, large ground-based telescopes can reach resolutions of a tenth of a second of arc or so. In the visible region, this is nearly as good as the Hubble Space Telescope, while in the near infrared, ground-based telescopes can equal the Hubble Space Telescope's performance.

Clearly a full atmospheric compensation system is complex and expensive. It is unlikely to be available to amateur astronomers. A significant improvement to the sharpness of an image can, however, be obtained at quite a modest cost by correcting just for the overall slope introduced to the wavefront by the atmosphere. That slope may be corrected by using a tip-tilt mirror. This is a plane mirror that can be tilted rapidly in any direction to compensate for the overall inclination of the wavefront. Commercially produced devices are now available for use on small telescopes.

Two or more telescopes can also achieve high resolution by being used together as an interferometer or as an aperture synthesis system (see the section on Radio Telescopes, below).

Radio Telescopes

The atmosphere is transparent only over two regions of the spectrum: the optical and radio regions (Figure 7.2 in Chapter 7). Although most amateur astronomers use visual telescopes, it is quite easy to build a small radio telescope to detect the brighter sources such as the Sun and the centre of the Galaxy. Large radio dishes, interferometers, and aperture synthesis systems seem likely though to remain the territory of professional astronomers and major observatories.

The huge dishes (Figure 1.23) that form most peoples' idea of a radio telescope are only a part of the whole instrument. The purpose of the dishes is to concentrate the radiation on to the detector, and also to shield the detector from unwanted radio emissions. The dishes are large partly because of the intrinsic faintness of radio sources. The unit used for intensity at radio wavelengths is the jansky,[1] defined as $1\ \mathrm{Jy} = 10^{-26}\ \mathrm{W\ m^{-2}\ Hz^{-1}}$, and some astronomical sources have millijansky intensities. The dishes are also large because the resolution of a telescope (Chapter 2) varies directly with the wavelength (Equation 2.3). A 10 m optical telescope thus has a resolution (ignoring atmospheric degradations) of $0.01''$, while a 10 m radio telescope operating at 21 cm wavelength would have a resolution of $5250''$, or nearly 1.5 degrees. Most radio dishes have similar optical principles to visual telescopes and operate either at prime focus, or as Cassegrain systems.

Once the radio waves have been brought to a focus, they must be converted to an electrical signal in order

---

[1] Named for Karl Jansky (1905–50) who, in 1932, started off radio astronomy by detecting radio emission from the Galactic centre.

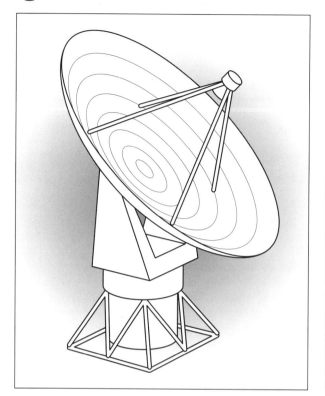

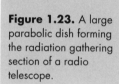

Figure 1.23. A large parabolic dish forming the radiation gathering section of a radio telescope.

to be detected. This is done by the feed. The feed is often the Yagi antenna that is well known as the ubiquitous television aerial, or a half-wave dipole. The latter is just the active element of the Yagi antenna and comprises two in-line conducting strips each a quarter of the operating wavelength long. At high frequencies, waveguides may need to be used and the radiation detected by Schottky diodes, superconducting tunnel junctions or bolometers. Details of these latter detectors are beyond the scope of this book and the interested reader is referred to sources listed in Appendix 2.

Once the radio wave has been converted to an electrical signal it is then amplified and processed by the radio receiver. Most receivers for radio telescopes operate on the same super-heterodyne principle used in domestic radios. Super-heterodyne receivers mix the main signal with a signal from a local oscillator at a similar but different frequency. The beat frequency between the two, usually known as the intermediate frequency, is then amplified. The receiver may also filter the signal and convert it to a voltage proportional to the input power. The final output is to a chart recorder, computer etc.

Interferometers

The resolution of even the largest radio dish in the world, the 300 m Arecibo telescope, is poor compared with that of the smallest optical telescope. Since it is impractical to build even larger single instruments, high resolution at radio wavelengths, and increasingly at optical wavelengths, is achieved through interferometry.

The basic operating principle of the interferometer is to combine the outputs from two or more telescopes when they are observing the same object. The outputs interfere with each other to an extent that depends upon the instantaneous path differences between the signals. As the object moves across the sky, those path differences change, and so the output from the interferometer oscillates (Figure 1.24). The form of the oscillation may then be used to provide high resolution data about the object (Chapter 2). The angular separation of two point sources may be easily determined using an interferometer, but more comprehensive imaging requires the use of advanced mathematics that is beyond the scope of this book. The interested reader is referred to sources listed in Appendix 2.

The resolution of an interferometer can be very high since it depends upon the separation of the two telescopes (Chapter 2), not the diameters of the telescopes. With very long base-line interferometry, that separation can be thousands of kilometres, leading to resolutions of milliarcseconds or less. However, there is no increase in sensitivity since no more light or radio waves are gathered.

The sensitivity as well as the resolution of a telescope with a diameter equal to the separation of the two telescopes making up the interferometer may be achieved through aperture synthesis. This technique was also first applied at radio wavelengths, but again is starting to be used in the infrared and visible regions. It relies upon the Earth's rotation to change the orientation in space of a pair of components of an interferometer through 360° over a 24 hour period. The two components thus trace out an annulus with a diameter equal to their separation and a thickness equal to their apertures over that period (Figure 1.25). If one component

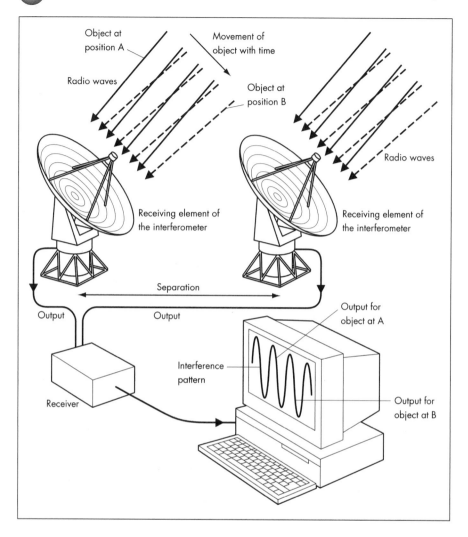

Figure 1.24. An interferometer and its output.

is then moved through its own diameter, the next annulus can be synthesised in the next 24 hour interval. Over a period of time, by moving the two components from being next to each other to their maximum separation, or vice versa, adjacent annuli can be synthesised to give the effect of observing with a normal telescope with an aperture equal to the maximum separation.

The technique can only be used on objects that will remain unchanged over the period of observation. The process may, however, be speeded-up by using many telescopes. For example, an interferometer with five telescopes could provide ten different separations

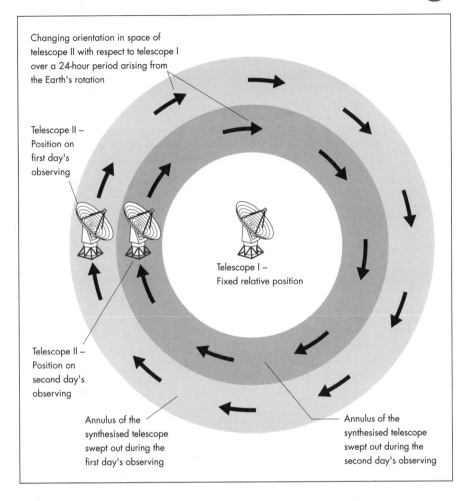

Changing orientation in space of telescope II with respect to telescope I over a 24-hour period arising from the Earth's rotation

Telescope II – Position on first day's observing

Telescope I – Fixed relative position

Telescope II – Position on second day's observing

Annulus of the synthesised telescope swept out during the first day's observing

Annulus of the synthesised telescope swept out during the second day's observing

Figure 1.25. Aperture synthesis, showing the first two annuli swept out during the first two days of observing. Further movement of telescope II through its own diameter during subsequent observing sessions will complete the synthesised telescope.

simultaneously (Figure 1.26), one with ten telescopes would provide 45 separations, etc. Also, in practice, only 12 hours of observation are needed since the other 12 hours can be reconstructed within the computer recording the observations. Even shorter periods of observation can be achieved by using several arms to the interferometer, oriented at different angles to each other. Thus the Very Large Array (VLA) in New Mexico, for example, has a Y-shaped configuration and twenty-seven 25 m diameter radio dishes with a maximum separation of 36 km.

If all the annuli required for the equivalent normal dish are synthesised, then it is a filled aperture system. If some annuli are omitted, then it is an unfilled aperture system. This latter is common for the larger baseline systems such as the UK's MERLIN (Multi-Element

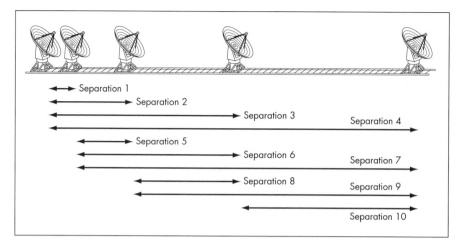

Figure 1.26. An aperture synthesis system with five telescopes can synthesise ten annuli simultaneously.

Radio Linked Interferometer Network), which has a maximum separation of 217 km. Unfilled aperture systems synthesise the resolution but not the complete sensitivity of the equivalent normal telescope.

Chapter 2

Telescope Optics

Point Sources

Telescopes normally use two lenses and/or mirrors to produce a magnified, and in the case of point sources, brighter image. There are many different designs, the basic properties of several of those more commonly encountered having been discussed in Chapter 1. In this chapter we take a closer look at the details of some of those designs.

Astronomical telescopes operate with the object at an effectively infinite distance. When used visually, by someone with normal eyesight, they produce a beam of parallel rays of light from the eyepiece, which is then focused by the eye to produce the observed image. The focal points of the objective and the eyepiece therefore coincide in normal use. It can easily be seen from Figure 2.1, that the angle for the exit rays (b) is related to the angle for the incoming rays (a) by the ratio of the focal lengths of the objective and the eyepiece. This ratio, for visual work, is the magnification of the telescope:

$$\text{Magnification} = M = \frac{b}{a} = \frac{f_o}{f_e}. \qquad (2.1)$$

For point sources, that is, images that are smaller than the detecting elements of the detector, i.e. the rod and cone cells in the case of the eye, the telescope also gives a brighter image. The situation for extended sources (images larger than the detector elements) is

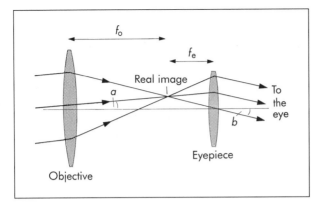

Figure 2.1. Optical paths in an astronomical telescope.

more complex and is considered later in this chapter. The increase in brightness for a point source is given by the light grasp, G, of the telescope, which is the ratio of the collecting area of the telescope to that of the pupil of the eye. The pupil of the dark-adapted eye is typically about 7 mm in diameter. The light grasp, ignoring any losses of light in the telescope, is therefore given by

$$\text{Light grasp} = G \approx \frac{\pi D^2 / 4}{\pi \times 0.007^2 / 4} \approx 20\,000 D^2 \quad (2.2)$$

where D is the diameter of the objective in metres. The Keck telescope ($D = 10$ m) thus has a light grasp of 2000 000.

The ultimate limitation on telescope performance is due to the wave nature of light. At any edge (such as the rim of an objective lens or mirror) the light waves will be diffracted (Figure 2.2). Even for a point source, therefore, in an otherwise perfect telescope, some of the light will be spread away from the centre of the image, and the actual image will not be a point. In such a case there will also be interference effects, and the image will consist of a circular bright central region, known as the Airy disc, surrounded by light and dark fringes (Figure 2.3). If the optics of a telescope are diffraction-limited, then its angular resolution, A, is given by the radius of the Airy disc:[2]

---

[2] Angular measure is normally in degrees (°), minutes of arc (′) and seconds of arc (″), with

$$360° = \text{a full circle}$$
$$60′ = 1°$$
$$60″ = 1′$$

Figure 2.2.
Diffraction of light at an aperture.

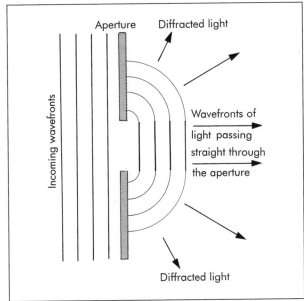

$$A = \frac{1.22\lambda}{D} \text{ radians} \qquad (2.3)$$

where λ is the operating wavelength and D is the diameter of the objective (in the same units).

Thus, for a telescope used visually, with $\lambda = 500$ nm, we have

$$A \approx \frac{0.122}{D} \text{ seconds of arc} \qquad (2.4)$$

where D is now in metres.

Since the resolution of the eye is typically 3–5 minutes of arc, if a telescope is to be used at its diffraction limit, then it must magnify the image sufficiently for $(A \times M)$ to exceed the eye's resolution. Thus a minimum magnification of about $1300D$ would appear to be needed for a telescope used visually to realise its potential resolution in practice. However, in reality, Earth-based telescopes have their resolutions

---

Footnote 2 – cont'd

For some purposes, though, radians need to be used, and

$$2\pi \text{ radians} = 360° = \text{a full circle}$$

so that

$$1 \text{ radian} = 57.2958° = 57° \, 17' \, 45'' \text{ and } 1° = 0.01745 \text{ radians}$$

The use of hours, minutes and seconds as an angular measure is discussed in Chapter 4.

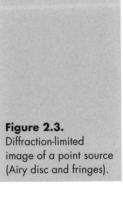

Figure 2.3. Diffraction-limited image of a point source (Airy disc and fringes).

limited by the atmosphere. From an average site, that resolution might be as poor as 2–10″. From a good site, the resolution might reach 1″ occasionally. The very best sites, such as Mauna Kea on Hawaii, with all possible precautions taken to minimise the effects of the dome and telescope, etc. can reach 0.25″. The Hubble Space Telescope, now that it performs to expectations, with its diameter of 2.5 m, thus betters our resolution of astronomical objects by a factor of about 5. Atmospheric compensation (Chapter 1) allows ground-based telescopes to reach resolutions of 0.1″ on occasion.

The eyepiece, since it consists of one or more lenses, produces an image of the objective, and this is known as the exit pupil (Figure 2.4). All the light passing through the objective must thus pass through the exit pupil, and as can be seen from the diagram, the exit pupil is also the point at which the emerging pencil of

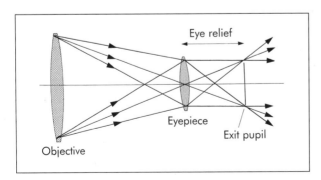

Figure 2.4. Exit pupil and eye relief of an eyepiece.

rays has its smallest diameter. The best place for the eye when observing with a telescope is therefore coincident with the exit pupil. For all the light to pass into the eye, the exit pupil must clearly be smaller than the eye pupil. The diameter of the exit pupil, D_{ep}, is given by

$$D_{ep} = \frac{Df_e}{f_o + f_e} \approx \frac{Df_e}{f_o} = \frac{D}{M} \tag{2.5}$$

(since $f_o \gg f_e$). The minimum magnification for a telescope, if all the light it gathers is to be used by the eye, is therefore given by

$$0.007 > D_{ep} = \frac{D}{M} \tag{2.6}$$

or

$$M_{min} \approx 140D. \tag{2.7}$$

The distance from the eyepiece to the exit pupil is known as the *eye relief*. For the simple lens shown in Figure 2.4, the eye relief is clearly approximately given by f_e. For comfortable viewing, its value should be about 6–10 mm. Practicable eyepiece designs (see below), using several lenses, may have values of eye relief from 2 to 20 mm.

The field of view of an eyepiece varies with its design, and can range from 30° to 70°, or more. The field of view of the telescope is then (if not restricted by other parts of the instrument) given by

$$\text{Field of view of telescope} = \frac{\text{Field of view of eyepiece}}{M}. \tag{2.8}$$

Extended Images

The light grasp of a telescope (Equation 2.2) shows that point sources are brighter when viewed through a telescope, and this is in line with most people's expectations of what a telescope does. However, the same is not true for extended sources, which in this context means any source whose image is larger than the size of the elements of the detector. Thus even stars at high magnifications and/or in poor atmospheric conditions can become extended sources. For extended sources, the greater amount of light gathered by the telescope must be spread over the larger area of the magnified image. Thus

$$\frac{\text{Surface brightness through the telescope}}{\text{Surface brightness to the eye}} = \frac{G}{M^2}. \quad (2.9)$$

Since from Equations (2.2) and (2.5), we have

$$G = \frac{D_o^2}{D_{eye}^2} \quad (2.10)$$

and

$$M = \frac{D_o}{D_{ep}} \quad (2.11)$$

we get

$$\frac{\text{Surface brightness through the telescope}}{\text{Surface brightness to the eye}} = \frac{D_{ep}^2}{D_{eye}^2}. \quad (2.12)$$

Now at the minimum magnification (Equation 2.7), we have

$$D_{ep} = D_{eye} \quad (2.13)$$

and so

$$\text{Surface brightness through the telescope} = \text{Surface brightness to the eye.} \quad (2.14)$$

Also at higher magnifications we have

$$D_{ep} < D_{eye} \quad (2.15)$$

and so

$$\text{Surface brightness through the telescope} < \text{Surface brightness to the eye.} \quad (2.16)$$

while at magnifications lower than the minimum (Equation 2.6)

$$D_{ep} < D_{eye} \quad (2.17)$$

so that not all the light from the telescope enters the eye. Effectively, in this case, the objective is reduced in size by the ratio D_{eye}/D_{ep}. The ratio of surface brightnesses therefore remains unity as the magnification is reduced below the minimum needed to fill the pupil of the eye. The surface brightness of an extended object is thus, at best, no brighter than when seen with the naked eye, and will usually be fainter than this. Other losses will ensure that in all circumstances the object is in fact fainter through a telescope than when seen directly!

This result is, to most people, very surprising, and perhaps contradicts their experience of using a telescope. Thus faint extended objects, like comets or the Andromeda galaxy, M31, do appear brighter through a telescope, and sometimes binoculars are advertised for night vision, whereas in fact their images cannot be brighter than those seen with the unaided eye. The answer to this paradox lies in the structure of the eye. The retina contains two types of detectors, cells known as rods and cones (from their shapes). The cones are of three varieties, sensitive to the red, yellow and blue parts of the spectrum respectively, and enable us to see in colour. The rods are sensitive to only the yellow part of the spectrum. In bright light the rods are almost depleted of their light-sensitive chemical (visual purple, or rhodopsin), and we therefore see via the cones. At low light levels, the rhodopsin in the rods regenerates slowly. When the rhodopsin is fully replaced, the rods are about one hundred times more sensitive than the cones. We then see principally via the rods in the retina. This behaviour results in two commonly experienced phenomena. Firstly, that of dark adaptation: our vision on going from, say, a brightly lit room into the dark slowly improves. This is partly because the pupil of the eye increases in size, but mostly because the regeneration of the rhodopsin is restoring the sensitivity of the rods. After about half an hour in dark conditions, one can often see quite clearly, whereas nothing at all had been visible to begin with. The second phenomenon is that of loss of colour in objects at night, and this occurs because the rods respond only to a single waveband of light. Now the rods and cones are not evenly distributed throughout the retina. The cones are concentrated in a region called the *fovea centralis,* which is the point where the image falls on to the retina when we look at it directly. The rods become more plentiful away from this region. If we look directly at a faint extended object, its image will fall on to the *fovea centralis* with its concentration of low sensitivity cones. However, on looking at the same image through a telescope, it will be magnified and some portions of it will have to fall on to parts of the retina away from the *fovea centralis,* and thus be detected by the higher sensitivity rods. Thus faint extended sources do appear brighter through a telescope, not because they are actually brighter, but because more sensitive parts of the eye are being used to detect them. This property of the eye can also be used deliberately to improve the detection of faint objects through the use of averted vision. In this, one deliberately looks a little way to the side of the object of interest. Its image then falls on to a region of the retina richer in rods, and becomes quite noticeably easier to see. Averted vision is a difficult trick to acquire, because as soon as the object flicks into view, one's normal reaction is to look directly at it again, and so it disappears. However, with practice the technique will become easier, and for any aspiring astronomer it is well worth the effort required to perfect it.

Objectives

The objective of a telescope can be either a lens or a mirror, or sometimes both. Its function is to gather the light and to focus it. The main properties required of both lenses and mirrors when used as objectives have been covered in Chapter 1, and there is no need to add to that discussion. Here, the practical requirements for objectives for telescopes in current use or construction are considered.

Except for Schmidt cameras and small telescopes for the amateur market (which have been discussed in Chapter 1), a telescope objective today is invariably a metal-on-glass mirror. Its surface is a paraboloid, or for the Ritchey–Chrétien design, a hyperboloid. The surface must be true to its correct shape to better than about one-eighth of its operating wavelength if it is to perform to its diffraction limit (Equation 2.3). For visual work this requirement is therefore about 60 nm. For some purposes, such as observations of the planets with their low levels of contrast, even more stringent limits may be necessary. The primary mirror of the Hubble Space Telescope deviated from its correct shape by 2 μm, which by everyday standards is a very tiny amount. However, it is over 30 times the deviation acceptable on even the cheapest of telescopes, hence the problems that the instrument has experienced.

The surface of the glass or other ceramic material used for the substrate of the mirror is still shaped by techniques that Newton and Herschel would recognise (Chapter 3). The mirror starts off as a disc of material called a blank. A second blank (called the tool) is placed beneath the mirror blank with some grinding powder, such as carborundum, between. The mirror blank is then moved back and forward over the tool in as random a motion as possible. With a coarse grade of grinding powder, the mirror blank quickly becomes concave and the tool convex. The process is continued until the mirror blank is hollowed out to the required depth. Sometimes this stage may be replaced by milling the surfaces to the required shapes with a diamond cutter in a numerically controlled machine. With the mirror blank of roughly the correct shape, the pits left by the grinding have to be removed. This is done by using a finer grade of grinding powder. The pits left by that powder then have to be removed with a yet finer grade of powder, and so on. Six or seven such stages may be needed before the mirror blank is smoothed sufficiently to be polished. Polishing is a similar process to grinding but using a softer powder. Usually ferrous oxide or cerium oxide powder is used for this process, and the tool is covered with a layer of pitch to provide a slightly malleable surface. When the mirror is finally polished, it then has to be figured. This is the process whereby the required shape for the surface is produced. It consists of a continued polishing of the surface, but with the polishing effect concentrated on to those areas that need to be reduced. Only a micron or so of material can be removed by figuring, and so the mirror must be close to its finally required shape before the process starts. The shape of the mirror must be tested during figuring to ensure that its shape is being altered towards that required. A variety of tests are available. One of the most straightforward is the Foucault test which, despite being simple enough for the equipment it needs to be constructed in a home workshop, can measure the shape of the mirror surface to a few tens of nanometres (more than accurate enough to have shown the problems for the Hubble Space Telescope mirror, at a cost of 0.00001% that of the COSTAR corrective optics!). The optical layout of the Foucault test is shown in Figure 2.5, and the shape of the surface is judged by the shadows appearing on the mirror as the knife-edge cuts the reflected beam. With large mirrors the tool may be considerably smaller than the mirror, and moved over the surface by a numerically controlled machine. With unusual shapes, such as the off-axis hyperboloids required for the segments of the Keck telescope, the blank may be

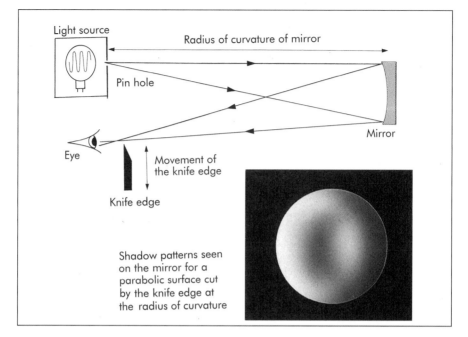

Light source

Radius of curvature of mirror

Pin hole

Mirror

Eye

Movement of the knife edge

Knife edge

Shadow patterns seen on the mirror for a parabolic surface cut by the knife edge at the radius of curvature

stressed during the figuring, so that it may be polished to a simple curve, and then it will relax to the more complex curve that is required, when those stresses are released.

The mirror, once correctly figured, must be sufficiently rigid to keep its shape under changing gravitational loadings as the telescope moves around the sky. With older telescopes, and with small telescopes, this was and is accomplished by making the mirror very thick. At larger sizes, however, the mirror then becomes very heavy, requiring a massive and expensive mounting. It will also have a large thermal inertia, and thus take a long time to match the local ambient temperature. While its temperature is changing it will have thermal stresses within it, which except for very low expansion materials, such as Zerodur and ULE (Ultra Low Expansion), will lead to distortions of the surface. The Russian 6 m telescope, for example, which has a Pyrex mirror, often gives poor quality images because the temperature of its mirror has not stabilised even by the end of the night! These problems can be partly overcome by making the mirror with a honeycomb back, as is the case with the 5 m Mount Palomar telescope, thus retaining most of the rigidity of a thick mirror, but reducing the weight and thermal inertia. More recently this has been extended to fabri-

Figure 2.5. Optical layout of the Foucault test and an example of the shadow patterns to be seen.

cating the mirror blank inside a rotating furnace. Two thin sheets of material are used for the front and back of the blank separated by numerous "struts" made from the same material. The whole is then fused together in the furnace, which because it is rotating, produces a dished front surface for the blank that is close to the finally required shape. An alternative approach is to make the mirror thin, but to mount it on active supports that are rapidly and frequently adjusted under computer control to distort the mirror so that its front surface remains in the required shape.

Eyepieces

Eyepieces matter very little for real astronomy.

Eyepieces are the most important accessories you will ever buy for your telescope.

These two extreme views of the importance of eyepieces summarise the confusion that reigns over this topic, for both are perfectly valid, and what is more, perfectly true *in certain circumstances*. It is the last caveat that is the important consideration when choosing among the enormous range of types of eyepiece available. The type of eyepiece that you will need depends upon the purpose for which you intend to use your telescope. Eyepieces are not cheap. At the time of writing one can pay over 15% of the cost of a complete 0.2 m Schmidt–Cassegrain telescope system for a single eyepiece! Since several eyepieces with a range of focal lengths will normally be needed, it is clear that they need to be chosen with care if money is not to be wasted.

The first quotation is applicable when the telescope will normally be used with a camera, CCD detector, photometer, etc. Then, the eyepiece is only needed for visually aligning the telescope on to the object to be studied before the real work begins. The performance of the eyepiece is almost immaterial; it can suffer badly from aberrations (see below) and still be quite adequate for its purpose. The same comment applies to an eyepiece to be used for guiding, though a somewhat higher quality will not go amiss in order to reduce eye strain, since it may be in use for considerable lengths of time. A guiding eyepiece will also normally need to be capable of being provided with illuminated cross-wires.

The second quotation applies when a telescope is to be used primarily for visual work. Even for this type of work, however, it is easy to waste money on expensive eyepieces. One of the major factors in the cost of an eyepiece is its field of view, and this can range from 30° at the bottom end of the range to 85° or more for ultra wide-angle eyepieces. Yet if we consider how a telescope is used in practice, a large field of view is rarely necessary. The object being studied will always be moved to the centre of the field of view for critical study whatever type of eyepiece is used, and on the optical axis eyepieces differ little in their performances. Thus the only applications that require eyepieces with a wide field of view will be searching for new comets and novae, and for "gawping" at large extended objects such as the Orion nebula. A wide field of view is useful when trying to find objects at the start of an observing session, but this can be achieved much more cheaply by using a low power eyepiece rather than a wide angle eyepiece (Equation 2.8). Another application that might appear to require a wide angle eyepiece is micrometry. This is a technique for measuring the angular sizes and separations of objects in the sky. The micrometer eyepiece has a fixed, centred set of cross-wires, and a second movable set whose position may be determined accurately. To measure the angular separation of a double star, for example, the fixed set of cross-wires would be set on to one star by moving the telescope, and the other set moved to the second star. However, it is more important to have an undistorted image for this purpose, and so the simpler designs of eyepiece are again to be preferred.

The second consideration in choosing eyepieces is their focal lengths, and hence the resulting magnifications of the telescope (Equation 2.1). It is usual to have eyepieces with a range of focal lengths to suit different observing conditions, and different objects. Four eyepieces, well chosen, will normally be sufficient for most purposes. As we have seen, there is a minimum magnification if all the light gathered by a telescope is to be utilised by the eye (Equation 2.7). This translates into a maximum focal length for an eyepiece for a particular telescope of

$$f_e \leq \frac{f_o}{140D} \qquad (2.18)$$

or a focal length of 70 mm and a magnification of ×30 for the widely found f10, 0.2 m Schmidt–Cassegrain telescope. Of course, a lower power eyepiece may still be used, and be of use when searching for objects, but some of the light gathered by the telescope will be wasted. The atmosphere will be sufficiently steady to allow telescopes less than about 0.25 m in diameter to perform at their diffraction limit a few times a year from most sites. There is then a minimum magnification of about 1300D if that limit is to be realised visually. This again translates into a maximum focal length, given by

$$f_e \leq \frac{f_o}{1300D}. \qquad (2.19)$$

For the same telescope as in the previous example, this would be a focal length of 8 mm and a magnification of ×250. High power, however, is more usually a disadvantage, since it reduces the contrast in the image, and often results in less detail being able to be seen. Under normal observing conditions, magnifications greater than ×200 would rarely be used. Thus for the f10, 0.2 m Schmidt–Cassegrain previously considered, a suitable range of eyepieces would have focal lengths of about 40 mm (×50), 25 mm (×80), 12.5 or 15 mm (×160 or ×130) and 6 mm (×330).

Other items to consider in choosing an eyepiece include the eye relief (Figure 2.4), which for comfortable viewing should lie between about 6 mm and 10 mm, and whether or not the eyepieces are parfocal. This last term means that, when pushed fully home into the eyepiece mount on the telescope, the eyepieces have their foci at the same point. Exchanging eyepieces therefore does not require the telescope to be refocused. Parfocal eyepieces will normally have to be bought as a set from a single manufacturer, and will be more expensive than equivalent non-parfocal eyepieces bought separately. However, the extra expense is usually worthwhile because of the increased ease of use of the telescope. Some designs of eyepiece may have up to eight lenses. Plain glass reflects about 5% of the light incident on to it, so with 16 surfaces, such an eyepiece could lose about 55% of the light entering it. It is thus essential to have the lenses in eyepieces covered with an anti-reflection coating to reduce the losses at each surface to 1% or less.

Finally, we come to a consideration of the choice of eyepiece optical design. There are a huge number of eyepiece designs, many of which are only slight variations on each other. Only the most commonly

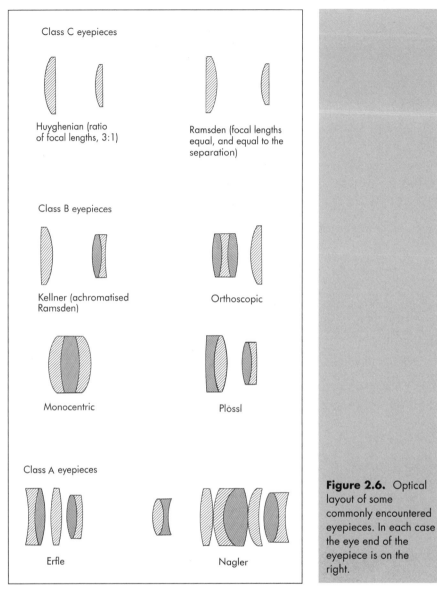

Class C eyepieces

Huyghenian (ratio of focal lengths, 3:1)

Ramsden (focal lengths equal, and equal to the separation)

Class B eyepieces

Kellner (achromatised Ramsden)

Orthoscopic

Monocentric

Plössl

Class A eyepieces

Erfle

Nagler

Figure 2.6. Optical layout of some commonly encountered eyepieces. In each case the eye end of the eyepiece is on the right.

encountered designs are considered here, and their optical layouts are shown in Figure 2.6. We may divide them into three groups on the basis of their cost and performance. At the bottom end we have class C designs such as the Ramsden and Huyghenian. These are not achromatic, and have small fields of view. Though cheap, they are generally not worth considering. In the middle range, class B, we have eyepieces such as the Kellner, Orthoscopic, Monocentric and

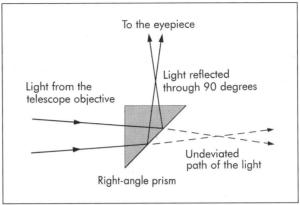

Figure 2.7. Star diagonal.

Plössl. These designs are well corrected for aberrations, have a wider field of view and are only a little more expensive than the class C eyepieces. They are the eyepieces of choice for most people for most purposes. The most expensive group, class A, are the wide-angle eyepieces, such as the Erfle and Nagler. As already discussed, these have few applications that justify their cost.

Accessories

Star Diagonal

The stellar diagonal is just a device to reflect the light from the telescope through 90° (Figure 2.7). It can make the observing position more comfortable when using a small telescope close to the zenith, but it will generally degrade the image to a noticeable extent. It is possible to purchase a diagonal in which several eyepieces can be mounted simultaneously. The eyepieces are in a rotating carousel, which enables them to be interchanged very rapidly. Parfocal eyepieces are almost essential for use in such a device.

Solar Diagonal

In the past, the reduction in intensity caused by reflection from plain glass has been used as the basis of a device for observing the Sun, known as a solar diagonal (Figure 2.8) or Herschel wedge. (Note that great care should always be exercised in solar observing.

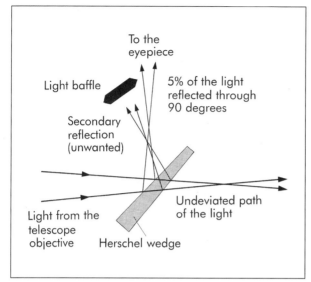

To the
eyepiece

Light baffle

5% of the light
reflected through
90 degrees

Secondary
reflection
(unwanted)

Undeviated path
of the light

Light from the
telescope
objective   Herschel wedge

Figure 2.8. Solar
diagonal (Herschel
wedge).

Make sure that you have read about and applied all the
precautions discussed in Chapter 8 before undertaking
any solar observation.) The reflecting surface for the
diagonal is just a plain piece of glass that diverts about
5% of the incoming light into the eyepiece. The glass is
wedge-shaped to prevent reflections from the second
surface from entering the eyepiece.

Solar diagonals are *not* now recommended for
observing the Sun. The reason for the change in the
recommendation on the use of solar diagonals is that,
in order to be safe, they must be used on a 50 mm
(2 inch) or smaller telescope, and at a minimum
magnification of ×300. This arises because plain glass
reflects about 5% of the incident radiation, and the
solar intensity must therefore be reduced by a further
factor of ×1700 in order for the final intensity of the
image to reach the recommended safe limit of 0.003%
of the unfiltered solar image. Since nowadays few
people use telescopes as small as 50 mm except as
finders, and rarely will seeing conditions in the
daytime be good enough to allow such high
magnifications, there is a great temptation to use the
solar diagonal on a telescope that is too large and with
too low a magnification. A solar diagonal on a 75 mm
(3 inch) telescope used at ×100 for example will
produce an image that is nearly 20 times brighter than
the safe limit.

Additional problems with the solar diagonal arise
from its use at the eyepiece end of the telescope. The

objective will therefore form a real image of the Sun inside the telescope and this may cause damage to the telescope structure. The wedge itself is "fail-safe" in that if it shatters due to the heat from the Sun, the eye is not exposed to the full solar brightness. However, even in normal use some 90% of the solar energy passes through the diagonal and emerges from the back of the telescope. It is easy to forget this when observing, and only to remember when the smell of scorching can no longer be ignored!

Full aperture filters consisting of aluminised Mylar film are better than solar diagonals since the light from the Sun is cut down before the telescope concentrates it (Chapter 8).

Barlow Lens

The Barlow lens is a diverging lens placed a short distance before the eyepiece. It serves to increase the effective focal length of the objective (Figure 2.9), and therefore (Equation 2.1) increase the magnification obtained from a given eyepiece.

Telecompressor

A telecompressor is the inverse of a Barlow lens. It is a positive lens placed before the eyepiece in order to decrease the effective focal length of the telescope. Its main use is to increase the field of view when imaging large objects.

Filters

Filters are used in photometry (Chapter 11) to define the wavelengths over which a star's energy is being

Figure 2.9. Barlow lens.

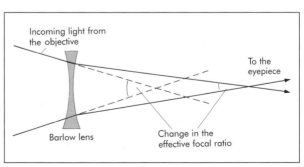

measured. They can also be used to enhance visual observing, and to improve CCD and photographic images.

The light pollution rejection filter is a filter designed to absorb strongly at those wavelengths where the light pollution is most intense. If the local street lighting is predominantly low pressure sodium lamps, then a light pollution rejection filter absorbing the region around 590 nm (the sodium D lines) can be very effective. Unfortunately high pressure sodium and mercury lamps emit light over most of the visible spectrum and so cannot so easily be eliminated.

The nebula filter is a narrow band filter designed to transmit the strong emission lines from gaseous nebulae. These lines are principally the forbidden lines of doubly ionised oxygen at 495.9 nm and 500.7 nm or the H-β line at 485.6 nm resulting in oxygen III and H-β filters respectively. Use of the filter enables longer exposures to be used, and hence images obtained with improved signal to noise ratios, since much of the light pollution is eliminated, but most of the light from the nebula retained.

The H-α cut-off filter is a filter that cuts out wavelengths shorter than about 650 nm, thus allowing through the H-α line at 656.28 nm, but eliminating most light pollution. It may be used for imaging gaseous nebulae. Note that it is not the same as the H-α narrow band filter and must *not* be used for solar observing.

The Comet filter is a filter that transmits light from about 460 nm to 550 nm so that the emission lines of molecular carbon (known as the Swan bands) are transmitted. Such a filter will help to enhance the views of the ion tail of a comet.

Aberrations

Those deviations of an image from perfection that are not due to diffraction (or grossly poor optics) are known as aberrations. We have already encountered the problems that early astronomers found with chromatic and spherical aberration, but those are not the only aberrations. There are six primary (Seidel) aberrations:

- spherical aberration
- coma

- astigmatism
- distortion
- field curvature
- chromatic aberration.

All, except the last, affect both lens and mirror systems. Chromatic aberration affects only lenses. There is not time for a detailed treatment of aberrations in this book, so just a brief summary of their effects will be given.

In spherical aberration, as we have seen (Figures 1.4 and 1.5), annuli of differing radii have differing focal lengths, while in chromatic aberration (Figure 1.6), light rays of differing wavelengths have differing focal lengths. In both cases, it is impossible to focus all the light rays simultaneously, and the image of a point source will consist of a bright centre surrounded by a halo of out-of-focus rays (Figure 2.10), these being coloured in the case of chromatic aberration. As we have seen, however (Figure 1.12), combining two lenses of different glasses can produce an achromatic lens in which the chromatic aberration is significantly reduced. The condition for achromatism in a cemented achromat, with two wavelengths λ_a and λ_b coinciding, is given by

$$\frac{R_1 + R_2}{R_1} \Delta n_c = \Delta n_f \qquad (2.20)$$

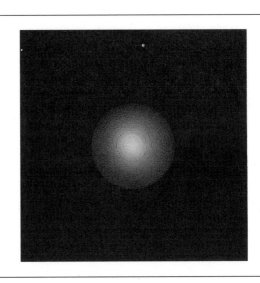

Figure 2.10. Image of a point source affected by spherical or chromatic aberration.

Figure 2.11. Image of a point source affected by coma.

where R_1 is the radius of the first surface; R_2 is the radius of the second and third surfaces, which are in contact (R_1 and R_2 are simply positive numbers – the usual sign convention of optics does not apply); Δn_c is the difference in refractive indices for the glass making up the first (usually crown glass) lens at wavelengths λ_a and λ_b; Δn_f is the difference in refractive indices for the glass making up the second (usually flint glass) lens at wavelengths λ_a and λ_b.

Coma is an effect whereby images of off-axis objects from different annuli of the lens or mirror are displaced by increasing amounts away from (or towards) the optical axis, and consist of rings of increasing sizes. The resulting image is a fuzzy blob of triangular shape pointing towards or away from the optical axis (Figure 2.11), and reminiscent of the image of a comet (hence the name). Astigmatism is differing focal lengths for rays in the vertical plane compared with rays in the horizontal plane (Figure 2.12). A cylindrical lens thus produces images with 100% astigmatism. The image of a point source from an ordinary lens or mirror with astigmatism will vary from a vertical line, through a uniform circle (known as the circle of least confusion, and the point of best focus), to a horizontal line (Figure 2.13). Distortion is differential transverse magnification for different distances of the image away from the optical axis. Decreasing magnification results in barrel dis-

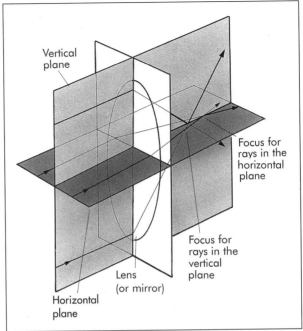

Figure 2.12. Astigmatism.

tortion, increasing magnification in pin-cushion distortion (Figure 2.14). Finally, field curvature has been encountered earlier in the discussion of the Schmidt camera (Chapter 1), and is caused because the focal plane is no longer a plane, but a curved surface. A flat detector such as a CCD or photographic emulsion will therefore not be in focus over its entire region.

Figure 2.13. Image of a point source affected by astigmatism at various points along the optical axis.

Image to one side of the position of the circle of least confusion

Circle of least confusion

Image to the other side of the position of the circle of least confusion

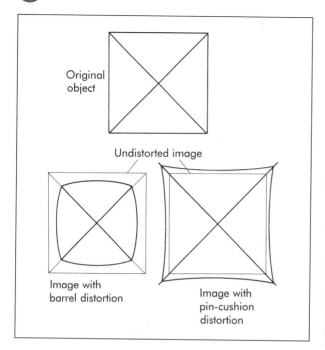

Original object

Undistorted image

Image with barrel distortion

Image with pin-cushion distortion

Figure 2.14. Image of an extended source affected by distortion.

Interferometers

Interferometers (Chapter 1) at both radio and optical wavelengths can provide much higher resolutions than single telescopes. A simple interferometer, however, does not produce a direct image, and aperture synthesis systems require a great deal of data processing in order to do so. Nonetheless, the basis of the high resolution of interferometers can be appreciated from considering the case of two close point sources.

The output from a two-element radio interferometer as it tracks a single point source across the sky is oscillatory (Figure 1.24). The equivalent for an optical interferometer is a series of fringes. If either type of interferometer should be observing two point sources close together in the sky, then each source will give rise to an oscillation or fringe pattern. When the path difference to the two elements of the interferometer from one source differs from that for the other source by a whole number of wavelengths, *plus* half a wavelength, a maximum from one fringe pattern will coincide with a minimum from the other. Assuming the sources are of equal brightness, the fringe pattern, or oscillations will then disappear. When the path difference to the two elements of the interferometer from one source differs from that from the other source by exactly a whole number of wavelengths, the maximum from one fringe pattern will coincide with the maximum for the other and the minimum from one with the minimum for the other. The fringe pattern or oscillations will then be doubled in intensity.

Thus when the two sources are perpendicular to the line of the interferometer, if the difference between their path differences is exactly half a wavelength, then the fringe pattern or oscillations will disappear, while for other positions in the sky it will be present to a greater or lesser extent. In that situation, we may see from the geometry in Figure 2.15 that

$$\sin\beta = \frac{(\lambda/2)}{S} \qquad (2.21)$$

or, since β is small and in radians

$$\beta = \frac{\lambda}{2S}. \qquad (2.22)$$

This is defined as the resolution of the interferometer. Comparing it with Equation 2.3 we may see that the resolution of an interferometer betters that of a conventional telescope with diameter equal to the separation of the elements of the interferometer by a factor of 2.44. The main improvement in resolution of an inter-

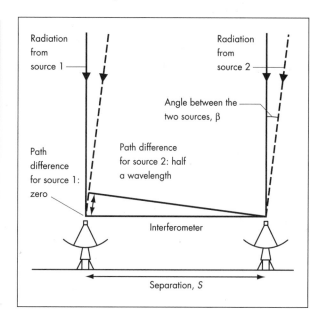

Radiation from source 1

Radiation from source 2

Angle between the two sources, β

Path difference for source 1: zero

Path difference for source 2: half a wavelength

Interferometer

Separation, S

Figure 2.15. An interferometer observing two point sources.

ferometer over a conventional telescope, though, arises from the ease with which large separations for the elements may be used.

The operation of an interferometer to measure the separation of two point sources requires the separation of the elements to be changed until the oscillations (radio interferometer) or fringes (optical interferometer) die away. Putting the value of that separation into Equation 2.22 then gives the separation of the point sources along the line parallel to the axis of the interferometer. A second interferometer perpendicular to the first can provide the separation parallel to that axis, and so the actual separation of the objects in the sky can be found. Alternatively an aperture synthesis system can produce the image directly after suitable data processing. Filled aperture synthesis systems though have the same resolution (Equation 2.3) as a conventional telescope but with a diameter equal to the separation of the elements of the interferometer.

Mountings

The function of a telescope mounting is to support the optical components, point them in the required direction and then to track the object as it moves. The mounting may be divided into two components: the telescope tube (which is often nothing like a tube), and the support for the tube (which we shall now call the mounting – see Chapter 6).

The tube in the case of a refractor is usually exactly that, with the lenses, supported at their edges, at the top and bottom of the tube. The main problems are the strains set up in large lenses by their weight, about which little can be done, and flexure in the lengthy tubes. The latter sometimes leads to tube designs based on joined frustra of cones.

For reflectors, the mirror can be supported at the back, as well as around the edges and, as previously discussed, this is one of the main reasons for the larger sizes of reflectors over refractors. The mounts for primary mirrors are often active, changing in response to the changing loads as the telescope orientation alters. In telescopes constructed three or more decades ago, complex mechanical linkages would adjust the mounting; nowadays, the adjustment is normally through computer controlled actuators (Figure 2.16).

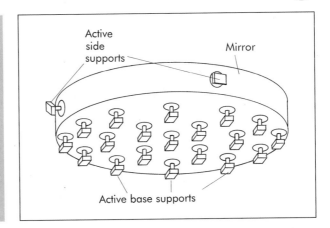

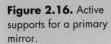
Figure 2.16. Active supports for a primary mirror.

The optics and their immediate supports in a reflector are then also mounted in a "tube" – the name is retained, even though the structure is often an open frame. Most such tubes are based upon the Serrurier truss (Figure 2.17) because this flexes in a known manner. Both the primary and secondary mirror supports flex as the telescope changes its orientation but, if designed correctly, the flexures are identical and so the optics remain in mutual alignment, even though the optical axis moves within the tube.

There are two main types of mountings for telescopes, called equatorial and alt-azimuth (alt-az) (Chapter 6), respectively. Examples are shown in

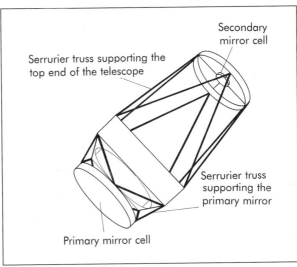

Figure 2.17. Reflector "tube" based upon Serrurier trusses.

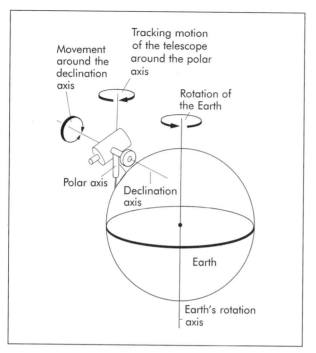

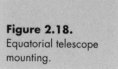

Figure 2.18.
Equatorial telescope mounting.

Figures 1.16 and 1.17. Although there are a number of different versions, the equatorial mounting always has one axis, the polar axis, parallel to the Earth's rotational axis. This is so that a single, constant-speed motor driving the telescope around this axis at one revolution per sidereal day in the opposite direction to the Earth's rotation will suffice to provide tracking (Figure 2.18). The other axis, at right angles to the polar axis, is called the declination axis. This system thus also has the advantage of giving simple direct readouts of hour angle (HA) or right ascension (RA) and declination (Dec) (Chapter 4). The main disadvantages of the equatorial mounting are that it is relatively expensive to construct, and the gravitational loads change in complex ways, making compensation for flexure difficult. The alt-az (which name derives from the allowed motions in *altitude* and *azimuth)* mounting has axes in the horizontal and vertical planes (Figure 2.19). This design is much cheaper to build, and simplifies the change in gravitational loading to just the vertical plane. In order to track an object in the sky, however, the telescope must be driven in both axes at varying speeds. Nowadays, with cheap computers,

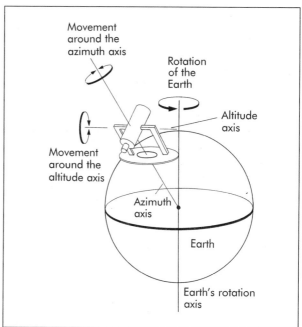

Figure 2.19. Alt-az telescope mounting.

this requirement is no problem, and so many of the recently constructed major instruments and also small commercially produced telescopes use alt-az mountings. One remaining problem is that the image rotates as the telescope tracks; the detector mounting must therefore also be rotated during any exposure, to produce sharp images.

A camera with a moderate focal length lens, such as might be used to image constellations, etc. may be very easily mounted so that it can track objects in the sky. The device is known as a barn door, Haig or Scotch mounting. It may simply and cheaply be made by any DIY enthusiast. It comprises two flat boards that are joined by a hinge along one of their edges. The hinge axis is aligned parallel to the Earth's rotation axis (i.e. on the north or north point in the sky). The camera is mounted on the upper board, and the tracking movement produced by a bolt through the lower board that bears on the upper board. The bolt is turned by hand at intervals of a few seconds. The amount and rate of motion required for the bolt will depend upon its thread angle and the distance from the hinge, and can be calculated or found by trial and error.

Observatories and Observing Sites

Even small telescopes are better when inside a purpose built observatory. This provides protection from the wind when the instrument is in use, and some shielding from unwanted lights. It also enables the telescope to be brought into action quickly so that advantage can be taken of brief clear spells. The classic design of a hemispherical roof with an open slot, rotating on a circular wall, has much to recommend it, but it needs to be built with precision if it is to function well. Small observatories are often made as DIY jobs, and therefore can come in many shapes and sizes. The design is only limited by the skill, imagination and facilities available. Medium sized observatories can also be home built, but may also be bought from specialist suppliers. In the latter case, the cost is likely at least to equal that of the telescope itself. The housing for a major telescope requires very careful planning and design, and is likely to form a significant fraction of the cost of the whole installation.

There is often little choice over the observing site, the observer's back garden often being the only possibility. Sometimes, however, it is possible to make a small telescope completely transportable so that it may be taken by car or van to a better site. Most major instruments constructed nowadays cluster together in a few sites, where the observing conditions are best. The principal requirements for a good site may be summarised as:

- away from light pollution;
- low dust content in the atmosphere, to reduce scattering;
- low water content in the atmosphere, to facilitate infrared observations;
- low diurnal temperature changes;
- steady atmosphere;
- height, to reduce the depth of the atmosphere above telescopes to a minimum;
- political stability of the host country;
- accessibility.

This long list limits suitable sites to a very few, such as Hawaii, La Silla, the Canary Isles, Kitt Peak, etc. As some measure of the determination of astronomers to get the best out of their telescopes, recently telescopes have begun to be operated from the Antarctic Plateau, where all conditions except the last are fulfilled!

Exercises

2.1 What is
   (a) the magnification (when used with an eyepiece of 25 mm focal length)
   (b) the resolution (in seconds of arc)
   (c) the light grasp of an f14, 0.2 m telescope?

2.2 What is
   (a) the minimum magnification for a 0.3 m telescope if no light from it is to be wasted?
   (b) the minimum magnification for a 0.3 m telescope if it is to be used visually at its diffraction-limited resolution?

2.3 Determine the resolution of the MERLIN system operating as an interferometer at a wavelength of 21 cm, given that 217 km separates the most widely spaced elements of the system.

Chapter 3

Modern Small Telescope Design

Introduction

The modern aspiring astronomer is faced with a bewildering choice of commercially produced telescopes, including all the designs considered in the preceding chapter. Yet only four decades ago the choice for a small telescope would have been between just a refractor and a Newtonian reflector. That change has come about because of the enormous interest that has grown in astronomy since the start of the space age and with the mind-boggling discoveries of the past 30 or 40 years. Except for some of the very small instruments which are unfortunately often heavily promoted in general mail order catalogues, camera shops and the like, the optical quality of these commercially produced telescopes is almost uniformly excellent. Although one product may be slightly better for some types of observation, or more suited to the personal circumstances of the observer, than another, most of them will provide excellent observing opportunities. The same general praise cannot be applied, however, to the mountings with which many of these telescopes are provided, and those problems are covered in Chapter 6.

The second problem associated with commercially produced instruments is that of cost. A new 0.2 m Schmidt–Cassegrain telescope can cost around a quarter to half the price of a small car. Other designs may be rather cheaper for the same aperture, but will nonetheless still represent a substantial outlay. One solution is to look at the second-hand market. The popular astronomy journals (Appendix 2) all have small advertisement sections, and telescopes can usually be found in there. Typically, the second-hand price of an instrument will be half its new price, and since telescopes have long lifetimes, this represents very good value for money. Indeed the advertisements often include phrases like "hardly used" or "only used twice", perhaps because the owners had not read Chapter 8 of this book first! There are few problems in buying a second-hand telescope, unless it has clearly been very severely treated. Mirrors can be realuminised, optics realigned, eyepieces replaced or the whole instrument refurbished by the original manufacturers or their agents at relatively small cost. The main points to watch are therefore concerned with the mounting and its drives: if these are worn or damaged then they can be very difficult to correct, especially with older instruments which may no longer be in production.

An alternative to second-hand instruments as a means of reducing costs, which is open to the reasonably skilled DIY enthusiast, is to purchase the optical components of the telescope and its tube, and then to construct a mounting. This may well be the preferred option anyway (Chapter 6) for telescopes with some of the more flimsy commercially produced mountings. The popular astronomy journals abound with articles giving detailed designs and constructional requirements for mountings from people who have chosen this route. No two such mountings are the same, of course, but the observer interested in such a project will find numerous ideas adaptable to his or her circumstances in only a few editions of these magazines.

The ultimate in astronomical DIY is to produce the optics for the telescope as well as the mounting. Indeed, for some people telescope making becomes more important than observing, and they produce highly sophisticated designs to far higher than commercially available standards. There is no need to go to such lengths however, and a simple telescope can be made from scratch that will give really excellent images. Assuming a certain degree of ingenuity on the part of the constructor, and depending upon the proportion of the materials that can be rescued from the scrap heap, a 0.2 m Newtonian telescope can be made for 5–20% of the cost of a similar sized commercially produced Schmidt–Cassegrain instrument. The author made just such a telescope at the age of 14, working largely from his own bedroom! The requirements of such a project are briefly outlined below, but the interested reader is referred to more specialised literature for details (Appendix 2).

Making Your Own

Although it is possible to make the optics for any design of telescope, in practice it is only the Newtonian reflector that is likely to be attempted by the vast majority of telescope makers, and this design should certainly be chosen for the first such project. The Newtonian requires only two mirrors, and the secondary is flat and available at low cost for purchasing. Only the primary mirror therefore needs to be produced, involving just a single concave paraboloidal surface. Generally speaking, when it comes to acquiring a telescope, the bigger the better! Size in this case refers to aperture, and not to length. However, the difficulties in making your own mirror scale as about the diameter cubed. The telescope maker should therefore be content to start with a 0.15 m or 0.2 m mirror, even though larger sizes might be financially possible.

The basic processes required to produce a mirror have been discussed in Chapter 2. For a 0.2 m mirror, the rough grinding would take about 10–20 hours, depending upon the depth required. It is therefore well worth getting the blank and tool diamond-milled to shape to start with, if this is possible. The cost of such preshaped blanks is usually not much more than that of unshaped blanks. For a small mirror such as this, Pyrex is adequate as a material for the mirror, and there is no need to use the much more expensive very low expansion materials such as Zerodur or ULE. Thereafter the constructor can expect to have to go through eight or so smoothing stages, at two to three hours per stage, polishing requiring 5–20 hours, and figuring requiring from five hours upwards. The total time required to produce a mirror is thus in the region of 40 to 60 hours, assuming that no mistakes are made. In addition, the telescope maker will also have to make a Foucault or other type of tester (Figure 2.5), and probably a stand on which the mirror may be worked. The latter stages of figuring are particularly time consuming, because the polishing process heats the mirror blank, and so it has to be left to cool down each time before it may be tested. Thus five or ten minutes actual work on the blank may take over an hour to accomplish.

The surface of the mirror should normally be within an eighth of a wavelength of light (0.000 000 06 m or 60 nm) of the correct paraboloidal shape if it is not to degrade images beyond the normal diffraction limit (Equation 2.4), and most specialist books on telescope making (Appendix 2) go to great lengths to describe how this may be achieved. However, for a first attempt, far poorer quality than this can be accepted. Deviations of a wavelength or more will still yield reasonable images in the telescope; the Hubble Space Telescope mirror, after all, is incorrect by two microns (four wavelengths) and was still able to be used even before the COSTAR correctors were added. The best strategy for a first-time telescope maker is therefore to produce a mirror of moderate quality, and to observe with it until its defects become a limitation, then to refigure it to a higher accuracy. Alternatively, a second higher quality mirror can be produced while the first is being used – a useful occupation for cloudy evenings!

Once the mirror has been shaped, it must have its reflective coating applied. A silver coating can be applied at home using some simple chemicals. Details of the process may be found in a reasonable chemistry textbook where it will be described as the silver mirror test. A silvered surface will, however, oxidise rapidly and will need reapplying at intervals of a few weeks; but this may be acceptable if the mirror is to be refigured after a short while, as suggested above. Normally, the reflective coating is aluminium, and this is often over-coated with a protective layer of transparent silicon oxide. The lifetime of such a coating, if the mirror is stored in dry conditions, is several years. However, both the aluminium and the over-coating have to be applied by evaporation on to the glass support inside a vacuum chamber, and the process is therefore unsuitable for undertaking at home. The cost of having a mirror professionally aluminised is not high, and suitable firms advertise in the popular journals, or local or national astronomy societies may have lists.

The secondary mirror for a Newtonian telescope is a flat set at 45° to the optical axis (Figure 1.10). It is quite small, and can be purchased fairly cheaply. Flat mirrors can be produced, if wished, by a variation of the techniques used for the primary mirror. However, unless a flat is already available, three mirrors have to

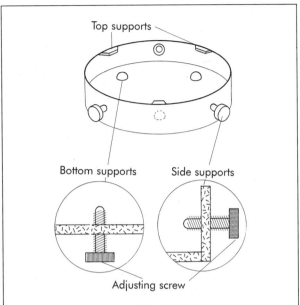

Figure 3.1. Primary mirror cell for a small mirror.

be produced at the same time, so that they may be tested against each other. For a first try, with a less than perfect primary mirror, the flat can be cut from a piece of plate glass (*not* ordinary window glass), and then silvered or aluminised. An elliptically shaped secondary obstructs the least amount of light, but again, for a first try, a rectangular one (much easier to cut) will only cause the loss of a small additional amount of light.

The mirrors, once produced, have to be supported in their correct relative positions and orientations within the telescope tube. The supports should be firm enough to prevent movement of the mirrors, and yet not apply sufficient pressure to cause stresses, which might distort the surfaces. For a 0.2 m primary, a mirror cell such as that shown in Figure 3.1 will be adequate. The weight of the mirror is taken on the bottom and side supports, and these may be adjusted to align the mirrors. The top restraints only just hold the mirror in place, for when the telescope is used at large zenith distances. A small secondary mirror may have a similar cell or be more simply attached to the supporting arms (often called the spider) using glue.

The lenses for eyepieces can be made if the telescope maker has access to a lathe; but several identical lenses have to be produced at the same time, and specialised equipment is needed, so that eyepieces are generally

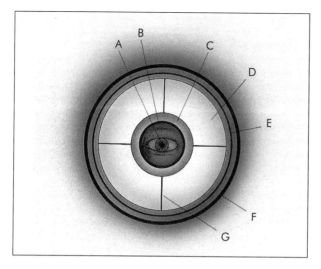

Figure 3.2. View through the eyepiece holder of a correctly collimated Newtonian telescope.
A: Eye reflected in the primary and the secondary (twice);
B: Eyepiece holder reflected in the primary and the secondary (twice);
C: Secondary reflected in the primary and the secondary;
D: Primary reflected in the secondary;
E: Secondary seen directly; F: Eyepiece holder seen directly;
G: Spider reflected in the primary and the secondary.

best purchased. Initial costs can be minimised by searching junk stores, where eyepieces from old microscopes and binoculars are often to be found. The eyepiece mount needs to hold the eyepiece, to allow interchange of eyepieces, and to enable the eyepiece position to be moved smoothly in and out for focusing. A good machinist can construct such a mount, but otherwise it will need to be purchased. Costs, however, can again be reduced by searching the second-hand market.

The mirrors in their cells or mounts and the eyepiece holder need supporting, and this is conventionally done by means of a tube. For a small telescope, plastic tubes can often be found of sufficient size and rigidity for this purpose. Alternatively, the tube can be constructed from wood or metal. A tube with a square cross-section is quite adequate, and is much easier to construct than one with a circular cross-section. Open tubes (Figure 2.17) can be used but have few advantages for small telescopes.

Once in the tube, the mirrors need aligning correctly, a process known as collimating the telescope. The eyepiece should be removed from its mount and the mirror positions and orientations adjusted until they have the appearance shown in Figure 3.2 when seen through the empty eyepiece mount, with all the components and their reflections concentric.

Finally, the tube and its optics need supporting on a mounting to enable them to be pointed at the sky and to follow objects as they move across the sky. Details of such mountings are to be found in Chapters 1, 2 and 6.

Commercially Produced Telescopes

Even the briefest of perusals of one of the popular astronomy magazines will reveal a bewildering array of advertisements for commercially produced telescopes, and some suppliers are also listed in Appendix 1. The optical qualities of these instruments are generally excellent, and the choice usually depends upon the available finance and the purpose for which the telescope is to be used. It goes almost without saying that anyone intending to purchase a telescope should not only read all the available literature carefully, but should also try out the instrument, preferably by finding someone else (perhaps in a local astronomy society) with the same telescope.

We may divide the vast majority of commercially available telescopes into four main groupings: Newtonian reflectors, refractors, Schmidt–Cassegrains and Maksutovs, and Dobsonian telescopes. Smallish Newtonian reflectors on equatorial mountings are probably the cheapest way to acquire a brand new telescope, and they will perform well, provided that they are kept collimated properly. Their main disadvantages are the ease with which the optics can become misaligned, their size and the general awkwardness of viewing with the eyepiece high on the side of the tube. Refractors such as those produced by Astro-physics Inc., Schmidt–Cassegrains such as the Celestron range, and Maksutovs such as the Questar, are very comparable in their performances. They all provide highly corrected images, and have closed tubes with rigidly mounted optics. The refractors, for a given aperture,

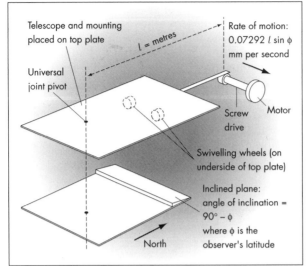

Telescope and mounting placed on top plate

l = metres

Universal joint pivot

Rate of motion: 0.07292 l sin φ mm per second

Screw drive

Motor

Swivelling wheels (on underside of top plate)

Inclined plane: angle of inclination = 90° – φ where φ is the observer's latitude

North

Figure 3.3. An equatorial platform for a Dobsonian telescope.

are generally lengthier, and therefore more awkward to transport, but have the advantages of not needing a secondary mirror to block some of the light and to (slightly – see Figure 8.4 in Chapter 8) reduce the image quality. If light grasp is your main consideration, then large aperture Newtonians on Dobsonian mountings, such as the Torus Optical range, are the best bet. These have the usual disadvantages of the Newtonian, coupled with the disadvantages of an alt-az mounting, but for the same money will provide 2–3 times the aperture (4–9 times the light grasp) of a refractor or Schmidt–Cassegrain, etc. It is possible to attach a Dobsonian-mounted telescope to a platform that can then be driven to track the stars for a short interval (Figure 3.3). Cassegrain and Ritchey–Chrétien tele-scopes are also available commercially, but generally only to special order and in quite large sizes, and are usually more suited to educational or small research applications.

Binoculars

Binoculars are just a pair of matched telescopes held together within a single framework that allows them to be pointed at and to focus on an object simultaneously. Both eyes may be thus be used for observing. Many people have difficulty in closing one eye, and so find binoculars easier to use than a single telescope. Binoculars provide an upright image through the use of internal prisms that also serve to shorten the length of the instrument. The performance of binoculars is usually specified using two numbers such as 7×30 or 10×50, etc. Here the first number is the magnification (i.e. $\times 7$ or $\times 10$), and the second the diameter of the objective lenses in millimetres (i.e. 30 mm or 50 mm). Recently gyro-stabilised binoculars have appeared on the market. These are much easier to use, since minor tremors are eliminated.

Aspiring astronomers are often advised to purchase binoculars instead of a telescope. While it is true that good quality binoculars can be purchased for the same cost as a cheap telescope, they are likely to disappoint if used for astronomical observing. This is because hand-held binoculars have too low a magnification to show very much in the sky. Even a magnification of $\times 10$, which is about the highest that can easily be hand-held, will not show lunar craters or the rings of Saturn. Higher magnifications are available, but can only be used if the binoculars are stabilised, or on some type of mounting. Then, however, they become as expensive as a good telescope.

Binoculars are also often advertised as "night vision glasses", implying that they provide a brighter image than that seen by the naked eye. However, as we have seen in Chapter 2 (Equation 2.16, etc.), this can never be the case for extended objects. Stars, though, will appear brighter through binoculars just as they do in telescopes.

Section 2

Positions and Motions

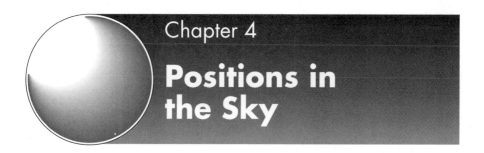

Chapter 4

Positions in the Sky

Spherical Polar Coordinates

Most people are familiar with the idea of plotting a graph. This is one example of a coordinate system, the *x* and *y* coordinates (abscissa and ordinate) providing a means of specifying the position of a point within the two-dimensional surface occupied by the graph. It is a simple extension of the idea to give completely the position of an object in space using three coordinates, *x*, *y* and *z* (Figure 4.1).

In astronomy there is frequently the need to specify the position of an object; however, neither the two or

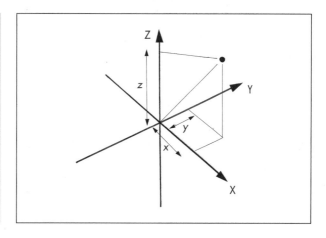

Figure 4.1.
Three-dimensional coordinate system based on orthogonal axes.

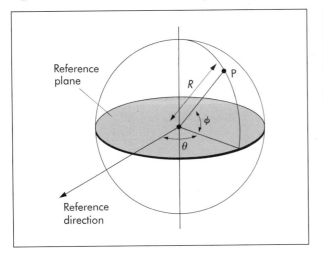

Reference plane

R

P

ϕ

θ

Reference direction

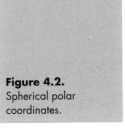

Figure 4.2.
Spherical polar coordinates.

three-dimensional Cartesian coordinate systems above are convenient. Instead, the three-dimensional position in space of an object is specified in a spherical polar coordinate system. This gives the position of an object, P, with respect to a point in space (the centre of the sphere) and a reference direction and plane, and results in coordinates, R, θ, ϕ (instead of x, y, z), where θ and ϕ are not linear distances, but angles (Figure 4.2). A familiar example of the use of spherical polar coordinates is to give positions on the Earth. Here, however, we normally only specify two of the three coordinates because the radial distance is almost constant at about 6370 km. Thus only the latitude and longitude need be given in order to fix a point on the surface of the Earth. Latitude is measured in degrees north or south of the equator, from 0° at the equator to 90° at the poles. Other systems of spherical polar coordinates may go from 0° to 180° (i.e. from south pole to north pole) for this coordinate. Longitude is measured in degrees from 0° to 180° east or west of the Greenwich meridian (Figure 4.3). In other systems, the equivalent coordinate may go from 0° to 360°. For the latitude and longitude position coordinate system, the reference plane is the Earth's equator, and the reference direction, the direction along the equator through the Greenwich meridian. Thus the old observatory at Greenwich is at a latitude of 51° 28′ N, and a longitude of 0°, while Mount Palomar observatory is at a latitude of 33° 21′ N, and a longitude of 116° 52′ W.

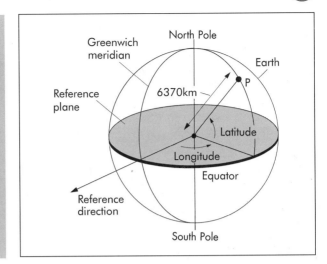

Figure 4.3. Latitude and longitude.

Celestial Sphere

Unlike objects on the surface of the Earth, those in the sky are distributed throughout the whole of three-dimensional space. Nonetheless we may still use a system of spherical polar coordinates analogous to latitude and longitude. We do this by imagining a huge sphere, centred on the Earth, and large enough to contain every object in the universe. We then project an object in three-dimensional space on to the surface of that sphere. Then just the two angular coordinates give its position (Figure 4.4). In other words, we ignore the radial coordinate of an object when it comes to assigning it a position in the sky. This accords with our common sense view of looking at the sky, when we are concerned with where to point a telescope, say, and the object's distance is immaterial for this purpose.

Any material object, such as a star, planet, galaxy, Sun, Moon, gas cloud, etc. can thus have its position in the sky represented by its projected position on the celestial sphere. The same, however, may also be done for less material objects. The north and south celestial poles (NCP and SCP, the points in the sky about which everything appears to rotate because of the Earth's counter-rotation), for example, can easily be added. These are the points where the Earth's rotational axis would intersect the celestial sphere if it were extended far enough. Thus we get the positions of the north and

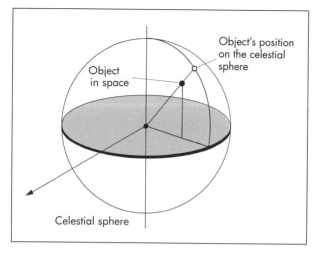

Figure 4.4. Celestial sphere.

south celestial poles in the sky (Figure 4.5). Most of the time, however, we can drop the qualifications "position of" and "celestial", and just talk about the north and south poles, etc. without any ambiguity. There is rarely likely to be any confusion with the real objects, and so we shall normally adopt this custom from now on. We may also add other items in a similar fashion. Thus we have the celestial equator. This we may obtain by imagining a line extending from the centre of the Earth in the plane of the Earth's equator. If that line is swept around, it will mark a great circle on the celestial sphere, which is where the plane of the Earth's equator meets the celestial sphere (Figure 4.5). Naturally this is

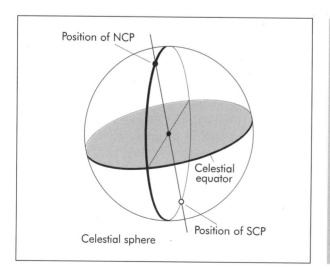

Figure 4.5. Celestial equator.

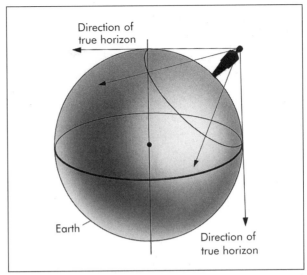

Figure 4.6. True horizon.

perpendicular to the line joining the poles. The zenith and nadir (the points in the sky directly overhead, and directly underfoot) for a particular observer can also be added. It might appear that we may plot in the horizon for that observer, in a similar manner to the way in which the celestial equator was obtained. Here, though, we normally deviate from strict accuracy because the true horizon will be affected by buildings, trees, hills, etc. and will also not be a plane because the observer's eyes are above ground level (Figure 4.6). Thus the celestial horizon is actually defined by the plane through the centre of the Earth perpendicular to the line joining the zenith and nadir (Figure 4.7). Conventionally, the horizon is shown horizontally, and the poles, zenith and nadir are shown on the outlining circle of the celestial sphere. Even though this is incorrect perspective, it simplifies geometrical calculations (Figure 4.8). The compass (cardinal) points on the horizon are called the north point (to distinguish it from the north pole), the east point, the south point (to distinguish it from the south pole) and the west point (Figure 4.8). Magnetic north (which is actually a magnetic south pole!) is displaced from true north by an amount that varies with place and time. Currently from the UK, the magnetic north point is about 3° to 4° west of true north, but this has changed from 10° east (in 1580) to 24° west (in 1800) in recent centuries. From the USA (except Alaska and Hawaii), the deviation (the

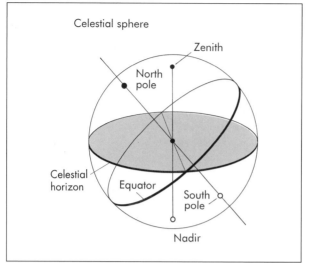

Celestial sphere

Zenith

North pole

Celestial horizon

Equator

South pole

Nadir

Figure 4.7. Celestial horizon.

angle between the north point and the magnetic north point) is currently about a degree or two either side of zero.

We can also plot out the yearly path of the Sun across the sky, which is known as the ecliptic, and along which lie the zodiacal constellations (Figure 4.9). The zodiac covers a band of the sky about 20° wide, centred on the ecliptic, within which are found the Sun, Moon and planets. Traditionally there are 12 constella-

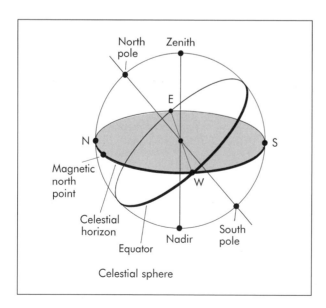

North pole   Zenith

E

N

S

Magnetic north point

W

Celestial horizon

Nadir   South pole

Equator

Celestial sphere

Figure 4.8. Conventional representation of Figure 4.7.

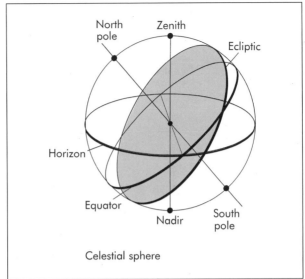

Celestial sphere

Figure 4.9. The ecliptic.

tions to be found within the zodiac, but the modern constellation boundaries place a thirteenth, Ophiuchus, into the region as well. The full list of zodiacal constellations is given in Table 4.1.

The conventional twelve zodiacal constellations are used in astrology to label twelve 30° wide bands (or signs) of the zodiac. However, precession (Chapter 5) means that these now have little correlation with the actual constellations. Thus, for example, the Sun is in the zodiacal sign of Sagittarius from 22 November to 21 December, an overlap of only two days with the true times that the Sun is within the constellation.

Table 4.1. The zodiacal constellations

Constellation	Dates of solar passage
Sagittarius	19 Dec to 21 Jan
Capricornus	22 Jan to 16 Feb
Aquarius	17 Feb to 12 Mar
Pisces	13 Mar to 8 Apr
Aries	19 Apr to 14 May
Taurus	15 May to 21 June
Gemini	22 June to 21 July
Cancer	22 July to 11 Aug
Leo	12 Aug to 17 Sept
Virgo	18 Sept to 31 Oct
Libra	1 Nov to 22 Nov
Scorpius	23 Nov to 30 Nov
Ophiuchus	1 Dec to 18 Dec

The motion of the Sun across the sky is of course actually due to the Earth's motion around its orbit. The ecliptic is therefore also the plane of the Earth's orbit marked on to the celestial sphere. We can imagine this being drawn by taking the line from the Sun to the Earth and extending it to the celestial sphere, and then making a mark as the Earth moves. More formally we may say that the ecliptic marks the intersection of the plane of the Earth's orbit extended outwards in all directions with the celestial sphere.

Altitude and Azimuth

We may now return to the problem of the positions of objects in the sky. The first of several spherical polar coordinate systems (with R ignored) that we shall examine is known as altitude and azimuth (abbreviated to alt-az). This system uses the horizon as the reference plane, and the north point as the reference direction. Altitude is then measured from 0° on the horizon to 90° at the zenith. Zenith distance is also used at times and is just (90° – altitude). The azimuth is the angle around the horizon from the north point to the great circle from the zenith passing through the object. It is measured from 0° to 180° east or west (Figure 4.10). Since the horizon and zenith are individual to each place, clearly the altitude and azimuth of the same object will, in general, be different as seen from different places.

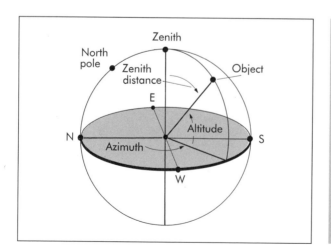

Figure 4.10. Altitude and azimuth coordinate system.

More importantly, the altitude and azimuth will change with time as the Earth rotates, and so cannot be used to identify a specific object in the sky except at a specific time.

Rotation

The change in altitude and azimuth of an object in the sky as seen by a particular observer is due to the movement of that object in the sky. Of course, this motion is actually due to the rotation of the Earth. For the discussions in this chapter, however, it is convenient to take the geocentric viewpoint and simply talk about the rotation of the sky or celestial sphere, while always remembering that it is actually the Earth that is rotating.

Thus the celestial sphere rotates on an axis running through the north and south poles once every 23 h 56 m 4 s. This period is shorter than the day because of the Earth's orbital motion. In the time that it takes the Earth to rotate through 360°, it has moved nearly a degree around its orbit. To return the Sun to its starting point as seen in the sky (the definition of the day), the Earth has to rotate that further fraction of a degree (Figure 4.11).

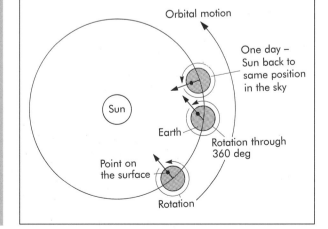

Figure 4.11. Rotation of the Earth (much exaggerated orbital motion).

Solar and Sidereal Days

The period of 24 hours is called the mean solar day, since it is based on the average time required to return the Sun to its starting place in the sky. The period of 23 h 56 m 4 s is called the sidereal day and is based on the time required to return a fixed object (star, galaxy, etc.) to its starting place in the sky. A clock that goes through 24 hours on its dial in 23 h 56 m 4 s is called a sidereal clock and it registers sidereal time. Two (24 hour) clocks, one keeping mean solar time and the other sidereal time, would agree with each other at the autumnal equinox (usually about 21 September, see later), and thereafter the sidereal clock would gain about 4 minutes a day. By the winter solstice (21 December), the sidereal clock would be 6 hours ahead, by the spring (or vernal) equinox (21 March) 12 hours ahead, by the summer solstice (21 June) 18 hours ahead, and by the next vernal equinox, the sidereal clock would have gained 24 hours and would again be in agreement with the mean solar clock. There are thus 366.25 sidereal days in a year, compared with 365.25 solar days.

Declination and Hour Angle

With altitude and azimuth as measures of the positions of objects in the sky, we found that they changed both with time and with the position of the observer, and that the manner of the change was quite complex. With the above definition of the sidereal day, we may now look at a new system of coordinates in which the changes occur in a simpler manner. This second system of coordinates is based upon the use of the celestial equator as the reference plane. The reference direction is obtained from the line in the plane of the equator that also goes through the prime meridian (Figure 4.12). The prime meridian is individual to each observer and is the great circle through the poles and the zenith (Figure 4.12).

The two angular coordinates are thus rather like latitude and longitude: up or down from the equator (dec-

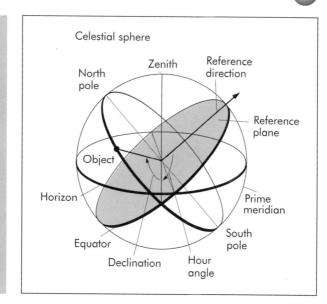

Celestial sphere

Zenith

Reference direction

North pole

Reference plane

Object

Horizon

Prime meridian

South pole

Equator

Declination

Hour angle

Figure 4.12.
Declination and Hour Angle.

lination, dec), and around from the prime meridian (hour angle, HA). This set of coordinates has the advantage over altitude and azimuth in that one coordinate for an object in the sky (declination) is now fixed, and does not change with position on the Earth, or with time. The other coordinate (hour angle), however, does still change, but in a simple manner compared with the changes in altitude and azimuth. Like latitude, declination is the angle from 0° to 90° north or south of the equator (usually indicated through the use of "+" for north, and "–" for south). The hour angle, though the equivalent of longitude, is different from it in two important ways. Firstly it is measured only westwards from the prime meridian rather than east or west from the Greenwich meridian. Secondly, the angular measure used is hours, minutes and seconds of time, not degrees, minutes and seconds of arc (for reasons that will shortly become apparent), with

$$1\,h = 15°$$
$$1\,m = 15'$$
$$1\,s = 15''.$$

Hour angle thus goes from 0 h to 24 h. For example, an hour angle of 6 h 45 m 20 s corresponds to the more normal angular measure of 101° 20', etc.

The value of the hour angle changes by increasing uniformly with time as the sky rotates. The HA is 0 h

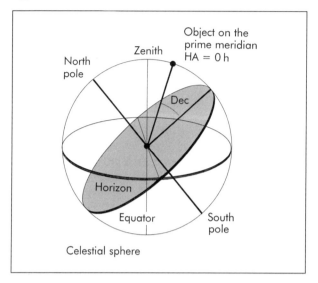

Celestial sphere

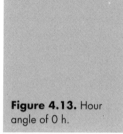

Figure 4.13. Hour angle of 0 h.

when the object is on the prime meridian (Figure 4.13). Then, provided that we measure time in sidereal units, one hour later, the HA will be 1 h. Two sidereal hours after meridian passage, the HA will be 2 h (Figure 4.14), and so on. Thus the reason for the units used for HA is through its direct relationship to the sidereal time elapsed since the meridian passage of the object.

The use of h, m, and s to measure HA clearly facilitates the calculation of the effect of time on its value. However, it complicates slightly the calculation of the effect of changing longitude, because the latter is mea-

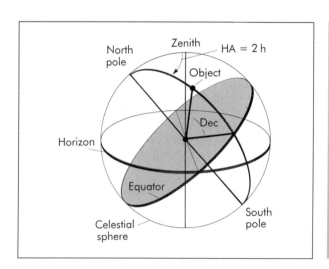

Figure 4.14. Hour angle of 2 h.

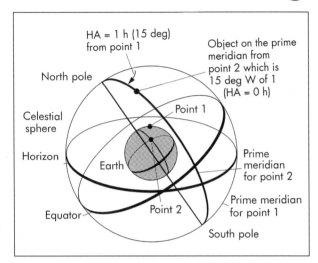

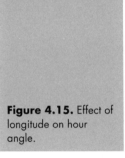

Figure 4.15. Effect of longitude on hour angle.

sures in degrees, minutes and seconds of arc, and east and west of the Greenwich meridian. Clearly if an object has an HA of 1 h from one place, however, it will be on the prime meridian at the same instant for a second place that is 15° to the west (= 1 h) of the first place (Figure 4.15). The relationship between HA and longitude is thus given by

$$HA_2 = HA_1 - \Delta Long(W) \qquad (4.1)$$
$$= HA_1 + \Delta Long(E) \qquad (4.2)$$

where HA_1 is the HA from point 1; HA_2 is the HA from point 2 at the same moment of time; $\Delta Long$ (W) is the difference in the longitudes between the two points, when the second point is west of the first, measured in h, m, s, etc.; $\Delta Long(E)$ is the difference in the longitudes between the two points, when the second point is east of the first, measured in h, m, s, etc.

Time

Having defined hour angle, we may now be more precise about our definitions of time.

Mean Solar Time

Mean solar time is defined as the hour angle of the mean Sun plus 12 h. The extra twelve hours is so that we get the conventional usage of 0 h being midnight, when the HAMS (hour angle of the mean Sun) is actually 12 h. The mean Sun is an imaginary body that moves around the equator (not the ecliptic) at a constant velocity that takes it through 360° in exactly one year. Mean solar time is thus a uniform measure of time, also known as civil time (see below), and is what we customarily use in normal life (except when it is adjusted for summer time, etc.).

Solar Time

Solar time is time by the actual Sun (as given by a sundial, etc.) and thus generally differs from mean solar time, because (a) the actual Sun moves around the ecliptic (not the equator), and (b) the actual Sun moves at a non-uniform velocity, because the Earth's orbit is elliptical, and thus the Earth's orbital motion, which produces the annual motion of the Sun across the sky, is non-uniform.

The difference between mean solar time and solar time is known as the equation of time (E):

$$E = \text{Solar time} - \text{Mean solar time.} \qquad (4.3)$$

The convention of adding 12 h also applies to solar time, and so this does not contribute to the difference. The value of E can be up to 16 minutes (Figure 4.16). The position of the Sun in the sky at civil midday (or any other fixed civil time, ignoring summer time), thus varies either side of the meridian. This variation, when combined with the Sun's motion in declination, results in a figure-of-eight shape known as the analemma, often to be found on antique maps and globes.

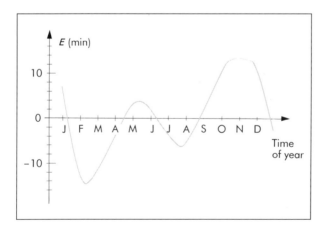

Figure 4.16.
Equation of time.

Civil Time

Sadly, astronomy's traditional role as guardian of the world's time keeping has now been usurped by atomic physics. The current unit of time is the second, which has been defined since 1967 as "the duration of 9192 631 770 periods of the radiation corresponding to the transition between the two hyperfine levels of the fundamental state of the atom caesium 133". The average results from a large number of atomic clocks around the world are used to provide "International Atomic Time" (TAI), which is used as the starting point for all other types of time scales.

On this basis, mean solar time as defined above is now known as Universal Time (UT), or Greenwich Mean Time (GMT). However, UT is still based upon the Earth's rotation and is therefore affected by changes in the rotation of the Earth such as the Chandler wobble.[3] The UT corrected for the Chandler wobble is called UT1, and is the basis of civil time keeping. It is kept to within 0.9 seconds of TAI by the occasional insertion or removal of a leap second.

Sidereal Time

Just as mean solar time was precisely defined via the hour angle of a particular point in the sky (determined by the imaginary mean Sun), so sidereal time can be defined in terms of the hour angle of a point in the sky. Any fixed point on the celestial sphere could be chosen to define sidereal time. The point actually chosen is known as the first point of Aries (FPA), or the vernal equinox. It is the point at which the equator and ecliptic intersect, and at which the Sun in its yearly motion passes from the southern hemisphere to the northern (Figure 4.17). Unlike the mean Sun, the FPA is fixed with respect to the sky (but see precession, Chapter 5), and so, as we have seen, the sidereal day is shorter by about 4 minutes than the mean solar day. The addition of 12 h to solar time to give 0 h at midnight means that sidereal and mean solar times agree at the autumnal equinox (about 21 September), instead of at the vernal

---

[3] A small irregular motion of the Earth's geographical poles, probably arising from movement of material deep inside the Earth.

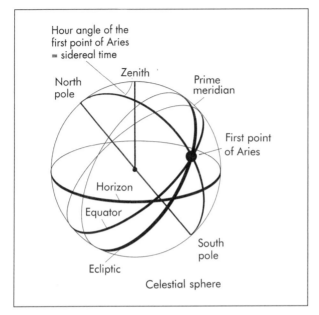

Hour angle of the
first point of Aries
= sidereal time

Zenith

North pole

Prime meridian

First point of Aries

Horizon

Equator

South pole

Ecliptic

Celestial sphere

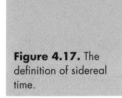

Figure 4.17. The definition of sidereal time.

equinox (about 21 March) when the HAMS and HAFPA (hour angle of the first point of Aries) are the same.

Just as was the case for mean solar time, sidereal time varies with longitude. From Equations (4.1) and (4.2), we may see that the sidereal time at a particular place (local sidereal time, LST) is given by

$$LST_2 = LST_1 - \Delta Long(W) \qquad (4.4)$$
$$= LST_1 + \Delta Long(E) \qquad (4.5)$$

where LST_1 is the local sidereal time from point 1 and LST_2 is the local sidereal time from point 2.

For sidereal time the main reference time (just as with mean solar time) is the LST at Greenwich, known as Greenwich sidereal time (GST). Equations (4.4) and (4.5) thus become

$$LST = GST - \Delta Long(W) \qquad (4.6)$$
$$= GST + \Delta Long(E) \qquad (4.7)$$

and GST is tabulated for midnight (GMT) in the *Astronomical Almanac* (Appendix 2) for every night of the year.

The calculation to find the LST for a particular place and at a particular civil time often seems to cause problems for the student, but is quite straightforward if it is approached calmly and logically:

1. Calculate the local mean time, LMT(0), corresponding to the previous midnight at Greenwich:

$$LMT(0) = 24\,h - T(W) \qquad (4.8)$$
$$= 0\,h + T(E) \qquad (4.9)$$

where $T(W)$ and $T(E)$ are the time zone corrections from Greenwich to the zone containing the place of interest, ignoring any summer time corrections, for places to the west and east of Greenwich respectively (e.g. Mount Palomar is 8 hours west of Greenwich so that $T(W) = 8$, and LMT(0) = 24 h − 8 h = 16 h = 4 p.m.).

2. Calculate the sidereal time interval, ΔLST, corresponding to the solar time interval, ΔLMT, from LMT(0) to the local mean time of interest, LMT(t). To avoid a common source of error, you should note that the numerical value of a sidereal time interval is always longer than that of a solar time interval. The one is obtained from the other by multiplying by the ratio of the number of sidereal days in a year to the number of solar days in a year:

$$\Delta LST = \frac{[LMT(t) - LMT(0)] \times 366.25}{365.25} = \Delta LMT \times 1.0027379.$$
$$(4.10)$$

3. Look up in a current copy of the *Astronomical Almanac*, the GST corresponding to the previous midnight, GST(0).

4. Calculate the GST at the local time of interest, GST(t), subtracting 24 from the result if it exceeds 24 h:

$$GST(t) = GST(0) + \Delta LST. \qquad (4.11)$$

5. Correct for the difference in longitude from Greenwich (Equations 4.6 and 4.7), to obtain the local sidereal time, LST(t), corresponding to the local mean time of interest. (Subtract or add 24 if the time goes over 24 h or under 0 h.)

Sample Calculation: To find the LST at 10.30 p.m. (LMT) on 16 November 1973 for the Mount Palomar Observatory.

Data:
Position   116° 21′ 30″ W (longitude)
            + 33° 21′ 22.4″ (latitude)
Time zone   8 h behind Greenwich

(1) LMT(0) = 24 h – 8 h = 16 h = 4 p.m.
(2) ΔLMT = 22 h 30 m –16 h = 6 h 30 m = 6.5 h
   ΔLST = 6.5 × 1.002 7379 = 6.517 7054 h
     = 6 h 31 m 4 s
(3) GST(0) = 3 h 39 m 42 s (from *Astronomical*
              Almanac for 1973)
(4) GST(t) = 3 h 39 m 42 s + 6 h 31 m 4 s
        = 10 h 10 m 46 s
(5) ΔLong(W) = 116° 21′ 30″ = 7 h 45 m 26 s W
   LST(t) = 10 h 10 m 46 s – 7 h 45 m 26 s
        = 2 h 25 m 20 s

Right Ascension and Declination

With the definition of a fixed point in the sky, the first point of Aries, we may return to the problem of the positions of objects in the sky, and now arrive at a much more satisfactory solution. We continue to use declination, for as we have seen, this is constant with time and position on the Earth. We replace hour angle, however, with a new coordinate, right ascension (RA), which is defined as the angle, measured in hours, minutes and seconds of time, around the equator from the first point of Aries in an easterly direction (Figure 4.18). This is in the opposite sense to the direction of measurement of hour angle. Since the first point

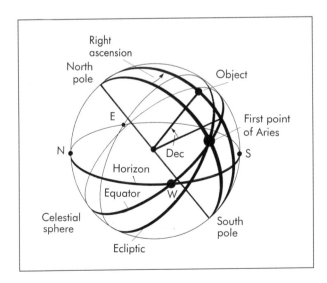

Figure 4.18. Right ascension and declination.

of Aries is fixed in the sky and rotates with it, the right ascension of another point fixed with respect to the sky is constant with time. Thus RA and Dec form a system of spherical polar coordinates for mapping the positions of objects in the sky. The equator is the reference plane and the direction of the first point of Aries is the reference direction. RA and Dec produce the commonly used coordinate system in astronomy, and most objects have their positions tabulated in terms of these coordinates.

For example, we have the positions of two well known stars:

Sirius (α CMa)   RA 6 h 45 m 9 s   Dec –16° 42′ 58″
Betelgeuse (α Ori) RA 5 h 55 m 10 s Dec +7° 24′ 26″

We may easily see (Figure 4.19) that there is a relationship between RA and HA for a specific object, the LST being given by

$$LST = RA + HA \qquad (4.12)$$

since LST is just the HA of the FPA.

This relationship (Equation 4.12) is fundamental to finding an object through a telescope. Most telescopes are mounted on equatorial mounts (Chapter 2), which have one axis parallel to the Earth's rotation, the other then being at right angles to the first. This system is chosen because it allows the telescope to be driven at a constant speed in one axis (one revolution per sidereal day about the polar axis in the opposite sense from the direction of the Earth's rotation) in order to counteract the rotation of the sky and to track an object being viewed in the telescope (Figure 2.18). It is therefore relatively cheap and simple to construct. Other mountings, such as the alt-az mounting (Figure 2.19) may be used, but these, though simpler to construct, require the telescope to be driven in both axes and at variable rates in order to track an object. Only recently, with the availability of cheap computers, have such mountings been used to any extent. With an equatorial mounting, rotating the telescope around the polar axis changes the HA at which the telescope is pointing, without changing the declination. Similarly, rotating the telescope around the declination axis changes the Dec at which the telescope is pointing, without changing the HA. Thus, if the axes are equipped with angle-measuring devices (usually called setting circles – essentially very large 360° protractors, see Chapter 6), the telescope can be set directly to the declination of the object. By going through the calcu-

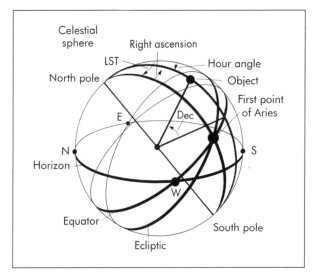

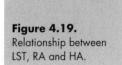

Figure 4.19.
Relationship between LST, RA and HA.

lation of HA of the object (as above) for the position of the observatory, for an instant say 10 minutes ahead of the current time, the telescope can then be set to that HA, and the telescope drive turned on when the correct time arrives. The object (given sufficiently accurate and well aligned axes and setting circles, and no mistakes(!)) will then be found centred in the field of view. Thus can faint objects be found which cannot be seen, except in the main telescope.

Most observatories will have a clock giving LST to enable this procedure to be carried out without having to go through the calculation every time. Many telescopes, including all larger instruments, automate the process to a greater or lesser extent, primarily by incorporating the LST into the telescope setting circles and read-outs to enable RA to be set directly. This of course makes life very much simpler; indeed, some advanced systems also contain catalogues of positions so that you just have to type in the name (e.g. Canopus, M31, Saturn, etc.) of the object you are interested in, for the computer to set the telescope for you!

Other Coordinate Systems

There are two other coordinate systems used to give the positions of objects in the sky, which you may encounter, and of which it is therefore worth being aware: celestial latitude and longitude, and galactic latitude and longitude.

Celestial latitude and longitude use the first point of Aries as a reference direction, and the ecliptic as a reference plane, compared with RA and Dec that use the first point of Aries and the equator. Galactic latitude and longitude use the direction towards the centre of the Galaxy (at RA 17h 45m 36s, Dec –28° 56′ 18″, in Sagittarius; the actual centre is now known to be about 4′ away from this point, but this position is still used for the galactic coordinates) as a reference direction, and the plane of the Galaxy (roughly the median line through the Milky Way) as the reference plane.

Heliocentric Time

For some purposes, such as observing variable stars, time as "seen" from the Sun is used in order to eliminate the varying travel time for the starlight in crossing different parts of the Earth's orbit. Any of the times discussed earlier may be converted to the heliocentric equivalent. The correction from the terrestrial time, in seconds, is given by

$$500\left[\sin\delta \sin\delta_{Sun} - \cos\delta \cos\delta_{Sun}\cos(\alpha_{Sun} - \alpha)\right] \quad (4.13)$$

where α and δ are the RA and Dec of the object being observed and α_{Sun} and δ_{Sun} are the RA and Dec of the Sun at the same moment.

Julian Date

Mention needs to be made of another convention used widely within astronomy for representing the time. This is based upon the solar day and is called the Julian date. It is used particularly when long time intervals are involved (for example when measurements of a visual binary star are being used to calculate its orbit) since it avoids the problems caused by the differing numbers of days in the months and of leap years.

The Julian date is a running day number that provides a simple way of calculating the time interval between two calendar dates. The Julian date starts at

Table 4.2. The Julian date

Date (midday, 1 Jan, Gregorian calendar)	Julian date
2050	2469 808.0
2025	2460 677.0
2000	2451 545.0
1975	2442 414.0
1950	2433 283.0
1925	2424 152.0
1900	2415 021.0
1850	2396 759.0
1800	2378 497.0
1750	2360 234.0
1700	2341 972.0
1650	2323 710.0
1600	2305 448.0
1200	2159 351.0
800	2013 254.0
400	1867 157.0
0	1721 060.0
400 BC	1574 963.0
800 BC	1428 866.0
1200 BC	1282 769.0
1600 BC	1136 672.0
2000 BC	990 575.0
4000 BC	260 090.0
4714 BC (24 Nov)	0.0

midday so that there is no change of date throughout the night, and began at midday on 1 January 4713 BC on the Julian calendar (24 November 4714 BC on the Gregorian calendar). Midday on 1 January 2000 saw the start of JD 2451 545 (see Table 4.2). Times other than midday are shown as decimal days. The Julian day number is the whole number part of the Julian date. The heliocentric Julian date is the Julian date corrected for the light travel time difference between the Earth and Sun. The running number of days throughout the year is shown in Table 4.3.

There is also a variation of the Julian date that starts at midnight on 17 November 1858 and called the modified Julian date (mJD). The modified Julian date is thus the Julian date minus 2400 000.5 days. It is sometimes used for data on spacecraft orbits.

Table 4.3. Running number of days throughout the year

Date (midday)	Normal year	Leap year
1 Feb	31	31
1 Mar	59	60
1 Apr	90	91
1 May	120	121
1 Jun	151	152
1 Jul	181	182
1 Aug	212	213
1 Sep	243	244
1 Oct	273	274
1 Nov	304	305
1 Dec	334	335
1 Jan	365	366

Spherical Trigonometry

Many people will have encountered trigonometry in some form or other as a means of dealing with geometry on flat or Euclidean surfaces. A triangle, for example, that has its internal angles designated A, B and C, and the lengths of the sides opposite to those angles designated by a, b and c (Figure 4.20), then has the following relationships between its internal angles and the lengths of its sides:

1 Sine rule

$$\frac{a}{\sin A} = \frac{b}{\sin B} = \frac{c}{\sin C}. \qquad (4.14)$$

2 Cosine rule

$$a^2 = b^2 + c^2 - 2bc \cos A. \qquad (4.15)$$

Since it is arbitrary which angles are labelled A, B or C, these relationships can be used to interrelate any angles and side lengths.

However, when we are dealing with positions in the sky, for example trying to find the angle between two objects or relating RA and Dec to altitude and azimuth, then we are concerned with the geometry that takes place on a spherical surface. This is the subject of spherical trigonometry, and it is very similar to Euclidean trigonometry. The two significant differ-

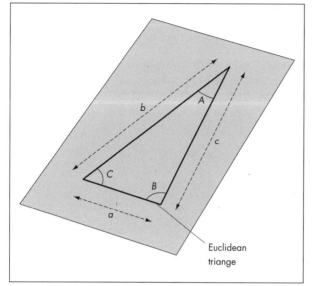

Euclidean
triange

Figure 4.20.
Euclidean triangle.

ences between Euclidean and spherical trigonometry are:

(a) The straight lines involved in the geometry occurring on a flat surface are replaced by great circles, or segments of great circles. (Note that spherical trigonometry does *not* deal with small circles.)

(b) The lengths of lines (i.e. segments of great circles) are given by the angle that they subtend at the centre of the sphere (Figure 4.21), *not* by their actual linear extents. The size of the sphere is thus not a concern of spherical trigonometry and its results apply just as much to figures drawn on the surface of a child's ball, as to figures on the surface of the Earth, or on the celestial sphere.

With these definitions, a triangle formed by the intersections of three great circles on a sphere (Figure 4.21), that has its internal angles designated A, B and C, and the angular lengths of the sides opposite to those angles designated by a, b and c, then has the following four relationships between its internal angles and the lengths of its sides. (Note that only three of the relationships are independent; the fourth can always be derived from the other three. Note also the similarity to the Euclidean relationships for the first two equations.)

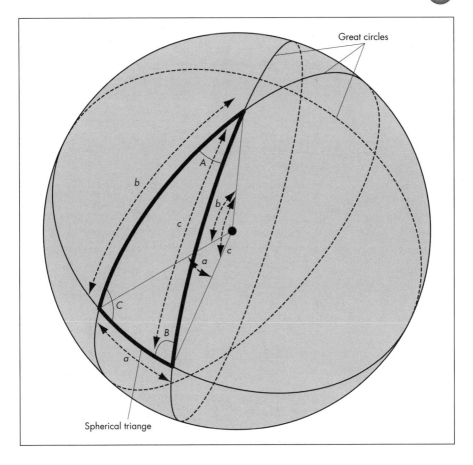

Great circles

A

b

b

c

c

a

C

B

a

Spherical triange

Figure 4.21.
Spherical triangle.

1 Sine rule

$$\frac{\sin a}{\sin A} = \frac{\sin b}{\sin B} = \frac{\sin c}{\sin C}. \qquad (4.16)$$

2 Cosine rule

$$\cos a = \cos b \cos c + \sin b \sin c \cos A. \qquad (4.17)$$

3 Four parts rule (since it involves four of the six angles and sides)

$$\cos a \cos C = \frac{\sin a}{\tan b} - \frac{\sin C}{\tan B}. \qquad (4.18)$$

4 Five parts rule (since it involves all the quantities except one of the spherical triangle's internal angles)

$$\sin a \cos B = \cos b \sin c - \sin b \cos c \cos A. \qquad (4.19)$$

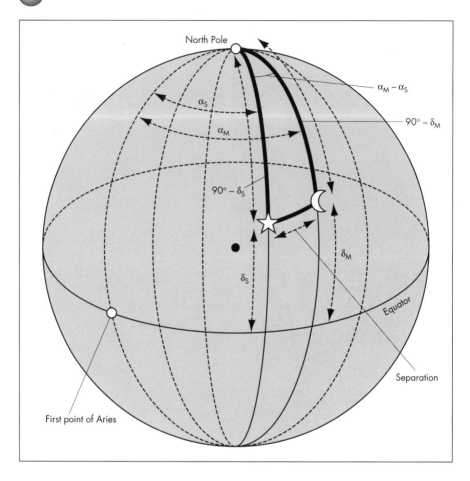

Figure 4.22.
Occultation.

Thus, for example, in order to determine if a star will be occulted by the Moon, we must show that its angular distance from the centre of the Moon is less than the Moon's angular radius. Given the RA and Dec of the Moon and star as α_M, δ_M, α_S and δ_S, we may use the spherical triangle between the celestial pole, the Moon and the star (Figure 4.22) since lines of constant RA are great circles. (Note that RA, HA, etc. must *always* be converted to degrees before inserting into trigonometrical formulae.) Hence

$A = \alpha_M - \alpha_S$
a = separation of the star from the centre of the Moon
$b = 90° - \delta_M$
$c = 90° - \delta_S.$

The cosine rule then gives us

$$\cos(\text{separation}) = \cos(90° - \delta_M)\cos(90° - \delta_S) \\ + \sin(90° - \delta_M)\sin(90° - \delta_S)\cos(\alpha_M - \alpha_S) \quad (4.20)$$

and for the actual values (where the star is Regulus, α Leonis)

$$\alpha_M = 10\text{ h }10\text{ m }10\text{ s} = 152°\,32'\,30''$$
$$\delta_M = +12°\,00'\,00''$$
$$\alpha_S = 10\text{h }08\text{ m }30\text{ s} = 152°\,07'\,30''$$
$$\delta_S = +11°\,58'\,20''$$

we get

$$
\begin{aligned}
\cos(\text{separation}) = {} & \cos(90° - 12°00'00'')\cos(90° - 11°58'20'') \\
& + \sin(90° - 12°00'00'')\sin(90° - 11°58'20'') \\
& \cos(152°32'30'' - 152°07'30'') \\
= {} & 0.999\,9094 \qquad\qquad\qquad\qquad (4.21)
\end{aligned}
$$

and so

$$\text{separation} = 46'17''. \qquad\qquad (4.22)$$

Since the lunar radius is about $15'$, this particular appulse will thus not give rise to an occultation. (An appulse is the instant when the separation of two moving objects passing by each other is smallest. The term 'conjunction' is often used for this moment, but strictly conjunction is the instant when the two objects have the same celestial longitude.)

Exercises

4.1 Convert the following angles into hours, minutes and seconds:

o	''	'
43	17	23
16	0	03
92	57	29
291	33	57

4.2 Convert the following angles into degrees (°), minutes (') and seconds ('') of arc:

h	m	s
03	06	05
14	46	0
18	09	28
21	13	13

4.3 What is the hour angle (in h, m, s) of an object 6 h 30 m of solar time after crossing the prime meridian?

4.4 If an object is on the prime meridian at the European Southern Observatory at La Silla, what is its HA from Greenwich? (La Silla (ESO): latitude 29° 15′ 26″ S; longitude 70° 43′ 48″ W.)

4.5 Calculate the local sidereal time at La Silla for 8 p.m. local time on 5 November, given that the previous GST(0) is 2 h 56 m 20 s. (La Silla time zone: 5 h west.)

4.6 What is the HA of (a) Sirius, (b) Betelgeuse, at 8 p.m. GMT on 5 November from La Silla (see Exercise 4.5, ignore precession)? Comment on whether either star would be visible. (Sirius: RA$_{(2000)}$ 6 h 45 m 09 s; Dec$_{(2000)}$ −16° 42′ 58″. Betelgeuse: RA$_{(2000)}$ 5 h 55 m 10 s; Dec$_{(2000)}$ + 7° 24′ 26″.)

4.7 (a) Use the cosine rule of spherical trigonometry to show that the altitude of an object is related to its RA and Dec (α and δ), the local sidereal time (LST) and the observer's latitude (ϕ) by;

$$\sin(\text{Altitude}) = \sin\phi \sin\delta + \cos\phi \cos\delta \cos(\text{LST} - \alpha)$$

(b) Use the five parts rule similarly to derive the formula for azimuth (hint: remember that $\sin^2 x + \cos^2 x = 1$):

$$\cos(\text{Azimuth}) = \frac{\cos\phi \sin\delta - \sin\phi \cos\delta \cos(\text{LST} - \alpha)}{\sqrt{1 - (\sin\phi \sin\delta + \cos\phi \cos\delta \cos(\text{LST} - \alpha))^2}}.$$

(c) Hence determine the altitude and azimuth at 05 h local sidereal time of an object at RA 3h and Dec +50° for an observer at a latitude of +60°.

Chapter 5
Movements of Objects in the Sky

Objects in the sky, including the so-called "fixed" stars, actually move in various ways, some quite complex, for a variety of reasons.

Diurnal Motion

Firstly, and obviously, the whole celestial sphere rotates in one sidereal day due to the counter-rotation of the Earth. Objects in the sky therefore move as seen from a fixed point on the Earth (Figure 5.1), and telescopes have to be driven around their polar axes to counteract this motion. An immediate consequence of

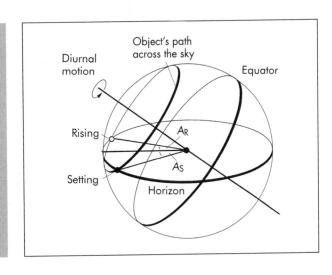

Figure 5.1. Diurnal motion.

this motion is that most objects in the sky are only visible for a fraction of the sidereal day between rising and setting. The hour angles of an object at rising and setting are related to its declination, δ, and to the latitude, ϕ, of the observer:

$$H_R = 24 - \cos^{-1}(-\tan\phi \tan\delta) \qquad (5.1)$$
$$H_S = \cos^{-1}(-\tan\phi \tan\delta) \qquad (5.2)$$

where H_R is the HA at rising and H_S is the HA at setting.

The right ascension of the object, α, gives the local sidereal times for its rising or setting using Equation (4.12). Reversing the calculation to obtain local sidereal time (Chapter 4) then gives the local civil time of the object's rising or setting. The position on the horizon of rising or setting can also be found. Its azimuth, A_R or A_S is given by:

$$A_R = \cos^{-1}(\sin\delta \sec\phi) \quad \text{east} \qquad (5.3)$$
$$A_S = \cos^{-1}(\sin\delta \sec\phi) \quad \text{west.} \qquad (5.4)$$

Circumpolar Objects

Equations (5.1)–(5.4) reveal that some stars never rise or set because the inverse cosine only has a meaning for values in the range ±1, and both $(\tan\phi \tan\delta)$ and $(\sin\delta \sec\phi)$ can lie outside this range. For example, with $\delta = 60°$ and $\phi = 52°$ (the latitude of the old Greenwich Observatory), we find $\sin\delta \sec\phi = 1.4067$ and so we cannot obtain the inverse cosine in order to find the rising or setting points. This is an example of a circumpolar object, or an object that is always above the horizon, and it is easily pictured (Figure 5.2). For northern hemisphere observers, the best known circumpolar object is the pole star, Polaris. Similarly, there are objects close to the south pole (for a northern observer) that never rise. The minimum declination for an object never to set is easily found (Figure 5.3):

$$\delta \geq 90° - \phi \qquad (5.5)$$

and similarly

$$\delta \leq \phi - 90° \qquad (5.6)$$

for an object never to rise.

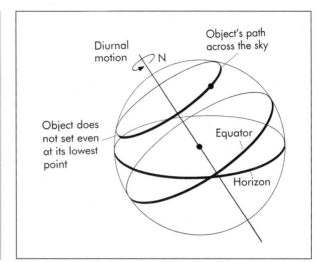

Figure 5.2.
Circumpolar object.

Seasons and Annual Motions

The Earth's orbital motion around the Sun, like its rotation, causes movements in the sky. For the more distant objects, the main effects are small and are dealt with later as aberration, precession and parallax. For objects within the solar system, however, the changes due to the Earth's orbital motion can be large. Those of the planets and the Moon are also dealt with later; here we just consider the effect on the Sun's position in the

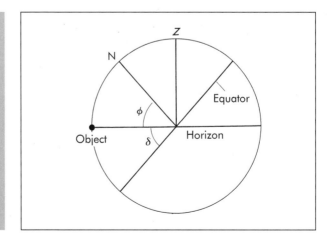

Figure 5.3.
Cross-section through the celestial sphere for an object that is just circumpolar.

sky. Since the Earth moves completely around the Sun in its orbit, the geocentric viewpoint is of the Sun moving completely around the sky, its path being called the ecliptic. This is not normally obvious, because when the Sun is in the sky, the other celestial reference points are invisible. Thus it is not easy to see that in early September, say, the Sun is to be found in the constellation of Leo, while by late September it is in Virgo. Instead of being aware directly of the Sun's movement, we are more aware of its consequences in the changing constellations visible at night and the seasons.

The constellations visible at night are clearly those opposite to the Sun in the sky, and so these change throughout the year. This change is reflected in names like the "Summer Triangle" for the brightest stars in Cygnus, Lyra and Aquila, visible on summer evenings. The solar movement is also marked at particular points along the ecliptic. One that we have already encountered is the first point of Aries, or vernal equinox (which is actually in Pisces; its position has moved owing to precession since it was first labelled by the early Greek astronomers – see later). The Sun passes through this point on or about 21 March each year. When the Sun is at the vernal equinox, it is also on the equator, and so we have days and nights of equal length. The other equinox, the autumnal, is opposite the first point of Aries in the sky, in Virgo, and the Sun passes through it on or about 21 September each year (Figure 5.4). The two other points commonly noted are the summer and winter solstices on or about 21 June and 21 December. These occur when the Sun is at its northernmost or northernmost points on the ecliptic; on the Gemini–Taurus border, and in Sagittarius, respectively. These points also mark the longest and shortest days, though because of the equation of time (Figure 4.16), the latest evening and the earliest morning are about 26 June and 16 June, and the earliest evening and latest morning about 12 December and 31 December. Our perception of day length, of course, is generally different from just sunrise to sunset, because of twilight. Twilight is due to light from the Sun reaching the surface after the Sun has set, as seen from that point, because it has been scattered in the Earth's atmosphere. Thus the sky remains bright after the Sun has set (or before it has risen). For astronomical purposes, twilight ceases when the Sun is 18° below the horizon. Civil twilight ends or begins when the Sun

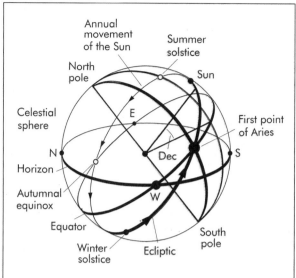

Figure 5.4. Annual movement of the Sun.

is 6° below the horizon, and nautical twilight when it is 12° below. At the height of summer, therefore, for any place north of latitude of 48.5° N (or south of a latitude of 48.5° S), twilight lasts throughout the night on the astronomical definition, and it never really gets dark. This period of "undark" nights increases as the latitude increases, until above 66.5° N, or below 66.5° S, the Arctic and Antarctic circles, there are occasions when the Sun becomes a circumpolar object.

This consideration of rising and setting times and points, and of the declinations of circumpolar objects, which we have just encountered, is correct in the absence of an atmosphere. The presence of the atmosphere, however, alters the calculations slightly, because refraction in the atmosphere causes objects to appear slightly closer to the zenith than their true positions. The effect is greatest for an object on the horizon, when it will appear to be about half a degree above the horizon. Sunrise and sunset will thus occur about 2 minutes earlier or later than predicted. For objects higher in the sky, the increase in altitude, R, is given approximately by

$$R \approx 58.2'' \tan z \qquad (5.7)$$

where z is the zenith distance of the object.

The other effect of the changing position of the Sun is that of the seasons, and this arises from the changing declination of the Sun in its path around the ecliptic,

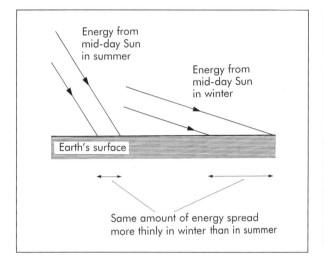

Energy from
mid-day Sun
in summer

Energy from
mid-day Sun
in winter

Earth's surface

Same amount of energy spread
more thinly in winter than in summer

Figure 5.5. Seasonal variation in the energy per unit area from the Sun.

not, as is commonly mistaken, from the changing distance of the Earth from the Sun as it moves around its elliptical orbit. There are three contributory factors to the larger amount of energy received from the Sun at a particular place in summer when compared with winter. The first is that the solar radiation is spread over a smaller area in summer than in winter (Figure 5.5), the second that the day length is longer in summer, and the third that the lower altitude of the Sun in winter results in more energy being lost by absorption in the atmosphere.

Movement of the Moon and Planets

Unlike the Sun, some objects in the sky can easily be seen to move against the background of the (relatively) fixed stars. Many of these objects have been known since antiquity; the planets from Mercury to Saturn, the Moon, plus the occasional comet, and their study probably represented the first stage of astronomy as a recognisable science. The motivation for their study lay in the fancied idea that they could foretell events on Earth, and coincidences between notable events such as battles, famines, deaths of kings, etc. and events in the sky would have seemed to support the possibility. Today, despite the plethora of astrology columns in popular newspapers, few intelligent people still hold that there is any such linkage. Instead the movements of the planets are studied in their own right for what they may reveal about gravity, the nature of the planets themselves and their satellites, plus, of course, the rather practical reason of knowing where they are in the sky so that they can be observed.

Moon

The Moon changes its position with respect to the background stars in a very short time, indeed its motion can be seen with the naked eye in an hour or so. This is due to the Moon's motion in its orbit around the Earth, though from an external viewpoint it would be truer to say that the Moon and Earth both have orbits around the Sun but with mutual perturbations, since the Moon's motion around the Sun is always concave inwards. The Moon is often said to keep the same face towards the Earth the whole time (Figure 5.6). This arises because the Moon's orbital and

Figure 5.6. Lunar rotation.

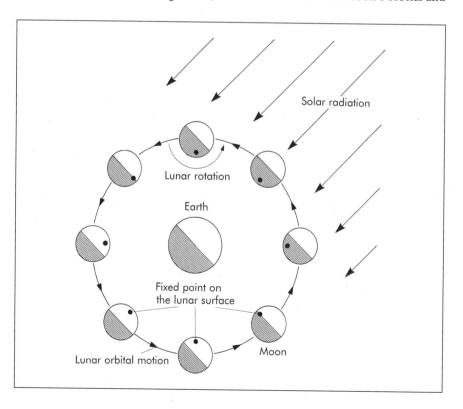

rotational periods are identical. That this is so is not a chance coincidence, but because in the past the tides produced in the Moon by the Earth have dissipated the Moon's rotational energy and slowed it down. The process only halts when the Moon has stopped rotating with respect to the Earth. Its rotation is then said to be tidally locked on to the Earth. The tides in the Moon have not disappeared when this has happened, but are now stationary with respect to the Moon, and so do not dissipate energy. The phases of the Moon occur from the changing proportion of its illuminated surface that we can see (Figure 5.16 below), and not, as is often mistakenly thought, from the Earth's shadow falling on to the Moon. The latter situation is a lunar eclipse (see later). The Moon's rotational motion is at a constant rate, but its orbital motion changes because the orbit is elliptical. Thus sometimes the Moon's rotation is ahead of its orbital motion, and sometimes behind, and we can see a little way "around the corner". This effect is called libration, and allows some 59% of the Moon's surface to be seen from the Earth.

Other Solar System Objects

Under this heading we include the major and minor planets and comets. All these objects move against the background stars at greater or lesser rates owing to the combined motions of the Earth and of the object around their orbits (Figure 5.7). Their RA and Dec have therefore to be calculated from their orbital parameters for a given moment of time, if they are to be found in a telescope. Fortunately this calculation has already been done for the principal objects and the results tabulated in, for example, the *Astronomical Almanac*. Likewise, the production of position predictions (an ephemeris) is the first urgent task of astronomers after the discovery of a new comet, so that it may be followed across the sky.

The paths of planets around the sky are complex (Figure 5.8), and their explanation caused many problems while the Earth was still regarded as the centre of the solar system. The main riddle was the way in which the normal motion of the planet across the sky from west to east (direct motion) would sometimes reverse (retrograde motion), then after a while the planet would again resume its usual direction of travel. The difficulties

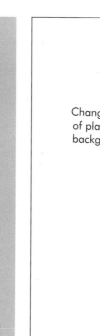

Figure 5.7.
Movement of the Earth
and other planets.

with understanding the motion of the planets, however, were not just due to the geocentric ideas, but also to the notion that the circle was the most perfect geometrical shape, and that therefore objects in the "perfect heavens" must move in circles. These two preconceptions eventually resulted in a model of the solar system described in AD 140 in the *Megale syntaxis tes astronomias* by Ptolemy (about AD 100–170). This book is better known as the *Almagest* (from the Arabic, *Al magiste*: the greatest) and had the planet moving around one circle (the epicycle), the centre of which in turn moved around a second circle (the deferent). To get good agreement

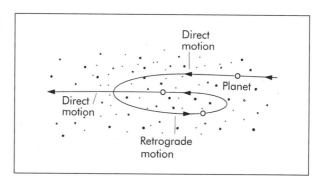

Figure 5.8. Motion
of a planet in the sky
(schematic).

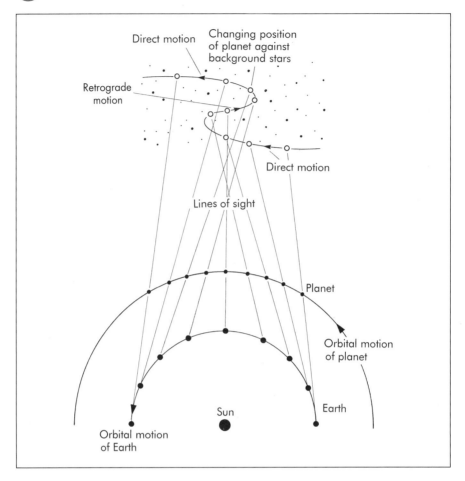

Figure 5.9.
Copernican
explanation of the
motion of the planets in
the sky.

with the observations, the centre of the deferent had to be displaced from the centre of the Earth, and in turn sometimes moved around a small circle. Although such constructions now seem outlandish, it should be remembered that Ptolemy's geocentric model of the solar system is actually the most successful scientific theory ever proposed, giving reasonably accurate predictions of the positions of the planets for some fourteen centuries. Eventually, of course, in 1543, the heliocentric model of the solar system was suggested by Nicolas Copernicus (1473–1543); but it was not until the publication of *Astronomia Nova* in 1609 by Johannes Kepler (1571–1630) that circular motions were replaced by elliptical orbits and thus finally solved the problem of the movements of the planets in the sky in terms of the relative motions of the Earth and the planets (Figure 5.9).

Proper Motion

We have so far regarded the stars for all practicable purposes as fixed in position. However, in many cases this is not quite true. Stars, including the Sun, are moving in orbits around the Galaxy with speeds typically of 200 to 300 km s^{-1}. Some stars are sufficiently close and/or have high enough velocities for this motion to lead to an observable change in their position. This change in position is called the proper motion of the star. Typical values for the proper motion are in the range 0.001$''$ yr^{-1} to 1$''$ yr^{-1}. The bottom limit is due to the degree of precision with which stellar positions can be measured, and is not a true cut-off. As higher precision positional measurements are obtained, for example, with the astrometric satellite *Hipparcos*, we may expect smaller proper motions to be determined. These values for typical proper motions do not lead to changes in the star patterns, etc. on human time scales, but on longer time scales, the constellations will change (Figure 5.10). The more distant stars and galaxies are effectively fixed in position except on time scales of hundreds of millions of years, though, in special cases, motions can be detected even for objects tens of millions of parsecs away. Thus very long base-line radio interferometry has revealed motions across the line of sight for relativistic jets in active galactic nuclei, and changes may be observed in some quasars owing to the motion of an intervening galaxy acting as a gravitational lens.

Precession

One of the first "motions" of the stars was noticed by Hipparchus of Nicaea (about 190–115 BC) in 130 BC. He noticed that Spica (among other stars) was some 2° further from the equinoctial points than had been the case 150 years earlier when observed by Timocharis. This "motion" turns out not to be a movement of the stars themselves, but of the reference point from which their positions are measured. As we have seen, that reference point is the first point of Aries, which we earlier took to be a fixed point on the celestial sphere. In fact, it is not fixed, but moves backwards around the ecliptic once every 25 700 years. The movement of the first

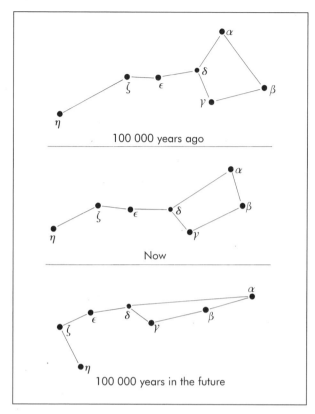

100 000 years ago

Now

100 000 years in the future

Figure 5.10.
Constellation of Ursa Major, 100 000 years ago, now, and in 100 000 years time.

point of Aries is due to the changing position in space of the equator, and this in turn arises because the Earth's rotation axis is swinging through space. Like a spinning top, the Earth's rotational axis swings (precesses) through space under the effect of the solar gravity on the Earth's equatorial bulge (Figure 5.11). The precessional motion causes the north and south rotational poles to trace out circles around the pole of the ecliptic (Figure 5.12). Thus it is just chance that we have a pole star (Polaris). As may be seen from Figure 5.12, no other bright star is approached by the pole in its motion anything like as closely as Polaris, and so for most of the 25 700 year cycle there is no pole star.

The effect of precession upon the position of an object is given by

$$\Delta\alpha = 3.4\Delta T(\cos\varepsilon + \sin\varepsilon\sin\alpha\tan\delta) \quad \text{seconds of time}$$
(5.8)

$$\Delta\delta = 50.4\Delta T\sin\varepsilon\cos\alpha \quad \text{seconds of arc}$$
(5.9)

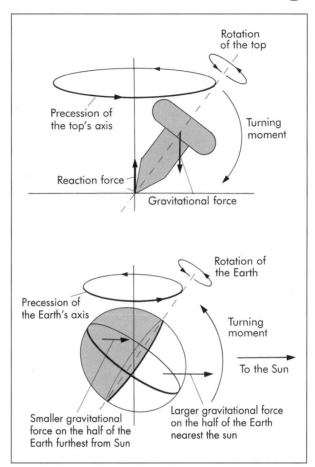

Figure 5.11.
Precession in a
spinning top and the
Earth (note that the
imbalance between the
gravitational forces on
the two halves of the
Earth has been much
exaggerated).

where $\Delta\alpha$ is the change in RA, $\Delta\delta$ is the change in declination, ΔT is the time interval between the epoch (the
time at which the position is tabulated) and the time
for which it is required in *years*, ε is the obliquity of the
ecliptic (23° 27′) and α and δ are the RA and Dec of the
object at the epoch.

Parallax

The Earth's orbital motion causes its position in space
to change by 300 000 000 km every 6 months. This is
sufficient to cause the nearer stars to shift their positions slightly owing to parallax (Figure 5.13). The
movement is small; only just over one second of arc
even for the nearest star (Proxima Cen), but it is

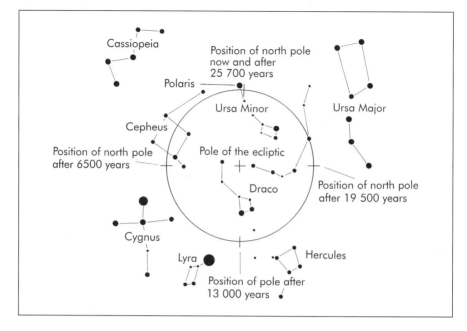

Cassiopeia

Position of north pole now and after 25 700 years

Polaris

Ursa Minor

Ursa Major

Cepheus

Position of north pole after 6500 years

Pole of the ecliptic

Draco

Position of north pole after 19 500 years

Cygnus

Lyra

Hercules

Position of pole after 13 000 years

sufficient to form the basic method of determining the distances of the stars:

$$D = \frac{1}{P} \qquad (5.10)$$

Figure 5.12. The movement of the north rotational pole due to precession.

where D is the distance in parsecs (1 pc = 3.3 light-years = 3×10^{16} m) and P is the parallax angle in seconds of arc (note that this is the half angle, not the whole parallax).

Over a year the object being observed will trace out a small ellipse centred on its position as seen from the Sun.

Aberration

This is the last of the motions of objects in the sky that we need to consider. Like parallax, aberration is an effect of the Earth's orbital motion. However, its magnitude is considerably larger than that of parallax: up to 20″, and it does not depend upon the distance of the object. It arises from the Earth's motion through space. Light travels at a finite speed (300 000 km s^{-1}), thus the Earth's motion carries the eyepiece (or detector, etc.) of a telescope along a short distance

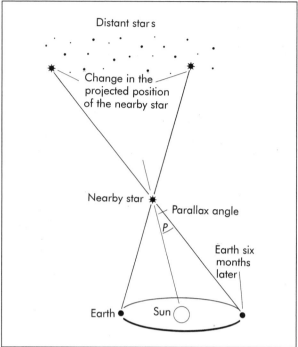

Figure 5.13. Parallax motion due to the Earth's orbital motion.

in the few nanoseconds it takes light to travel down the telescope (Figure 5.14). The observed position of the object is hence shifted towards the apex of the Earth's velocity at that instant of observation by a

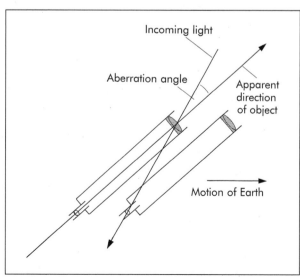

Figure 5.14. Aberration.

small amount. (The apex is the point in the sky towards which the Earth's orbital motion is directed at any given instant. It lies on the ecliptic and follows the Sun over a period of a year. At aphelion and perihelion it is exactly 90 degrees behind the Sun; at other times it gains or loses slightly on the Sun due to the ellipticity of the Earth's orbit.) Since the Earth's velocity changes direction around the orbit, an object observed throughout a whole year would move in a small ellipse centred on the position the object would have if the Earth were stationary. The aberration ellipse may be distinguished from the parallax ellipse because the instantaneous shift of the object is towards the position of the Sun in the case of parallax, and towards a point on the ecliptic 90° behind the position of the Sun in the case of aberration.

Relative Planetary Positions

Position with Respect to the Earth

The movements of the Earth and planets mean that their relative positions in space are continually changing. Several relative positions of the Earth, Sun and another planet are sufficiently important to be given specific labels (Figure 5.15). The most useful of these terms are opposition, when an outer planet may most easily be observed, and greatest elongation, which is the optimum time for observation of an inner planet. The best time for observing an inner planet is not inferior conjunction as might be expected, when the planet is at its closest to us, because we can then only see the unilluminated half of the planet. The relative positions result in a greater or lesser extent of the planet's surface being seen as illuminated. For the Moon, this results in the familiar phases (Figure 5.16). Mercury and Venus go through similar phases, and also noticeably change their angular sizes (Figure 5.17). The outer planets show phases, but they are always gibbous, and normally the effect is unnoticeable except, at times, for Mars.

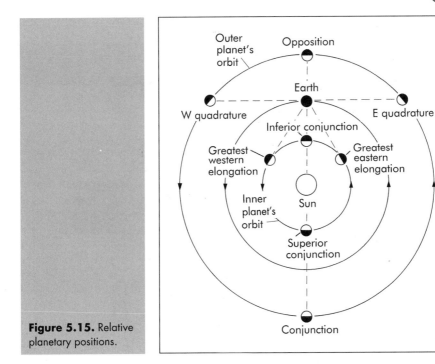

Figure 5.15. Relative planetary positions.

Eclipses

Some relative positions of the Earth and other solar system objects are of particular scientific significance, and these are eclipses, occultations and transits. Sometimes the relative orbital motions of the Earth and two other objects cause one of those objects to pass in front or behind the other. An eclipse occurs when the two objects are of comparable angular size, for example, the Sun and Moon, two of Jupiter's Galilean satellites, or two stars in a binary system. An occultation occurs when an object of much larger angular size passes in front of one that is angularly much smaller, such as the Moon in front of a star. A transit occurs when an object of much smaller angular size passes in front of one that is angularly much larger, for example, Mercury or Venus in front of the Sun, or a Galilean satellite in front of Jupiter. All these phenomena are of great interest because they may reveal details of the objects not otherwise easily observable, such as the solar corona or the rings of Uranus (discovered during observations of an occultation of a star by Uranus). Near approaches to mutual alignments result in con-

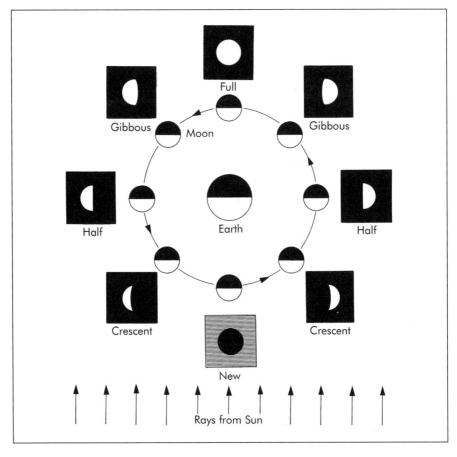

Figure 5.16. Phases of the Moon.

junctions and appulses when two planets, say, may come within a degree or two of each other in the sky, but these events, like quadrature, are of only aesthetic interest.

A solar eclipse occurs when the Moon passes in front of the Sun as seen from the Earth (Figure 5.18). By accident, their angular sizes are almost equal. Most of the time, the angular size of the Moon is slightly larger than that of the Sun, but at apogee (see below) its angular size is slightly smaller, and this results in annular eclipses. When the Moon passes close to centrally across the Sun, we get a total or annular eclipse; if the Moon does not pass close to centrally across the Sun, then we get a partial eclipse. Because the plane of the Moon's orbit is inclined to the ecliptic by about 5° (Figure 5.19) and because that plane rotates in space due to terrestrial and solar perturbations, the number of solar eclipses can vary from two to five per year.

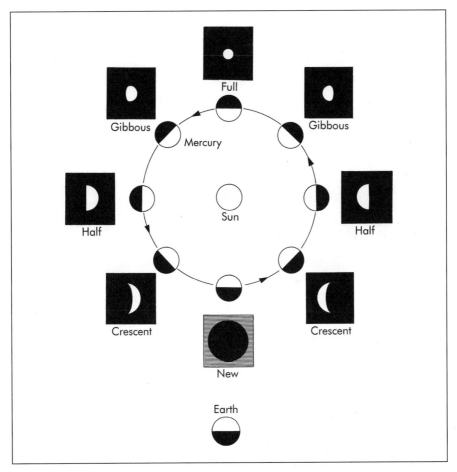

Figure 5.17. Phases of Mercury and Venus.

Lunar eclipses are quite different from all the other phenomena discussed in this section. For a lunar eclipse to be comparable with the other eclipses, it would have to be viewed by an astronaut on the surface of the Moon – and should be called a solar eclipse by the Earth! A lunar eclipse occurs when the Moon passes into the shadow cast by the Earth (Figure 5.20). For the same reason, therefore, as solar eclipses, lunar eclipses can only occur when the line of nodes of the Moon's orbit passes close to the Sun (Figure 5.19). There can thus be up to three lunar eclipses per year. Lunar eclipses are much rarer than solar eclipses; however, they appear to be more common because they can be seen from half the Earth at a time.

Transits occur for Venus and Mercury against the Sun, and for planetary satellites against the parent planet. They are no longer very significant, but in the

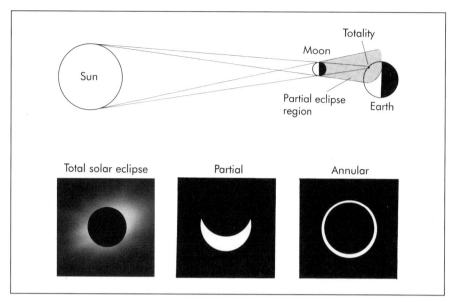

past they were important as a means of measuring the astronomical unit, and for Ole Rómer's (1644–1710) determination of the velocity of light by timing the transits of the Galilean satellites. In an occultation, the angularly smaller body disappears behind the angularly larger body. Lunar occultations are of use in identifying close double stars, and measuring stellar diameters. Occultations by planets can give information on the outer parts of the planets' atmospheres, since the light from the star can still be detected even

Figure 5.18. Solar eclipse.

Figure 5.19. Limits on the occurrence of eclipses.

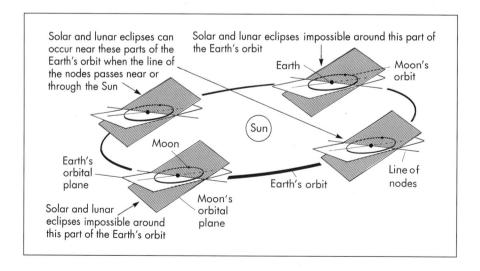

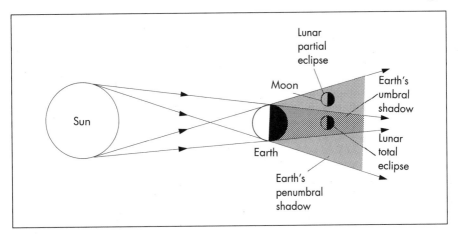

Figure 5.20. Lunar eclipse.

when passing through an appreciable amount of material.

Position in an Orbit

In an elliptical orbit of one object around another, the separation of the two objects varies; this also leads to certain relative positions being identified by name. The maximum and minimum separation points are called the apsides (singular apsis), and the line joining them, the line of apsides, or more commonly, the major axis of the orbit. The maximum separation is signified by the prefix "Ap", the minimum by the prefix "Peri", followed by a suitable signifier for the object involved. Thus for an object in orbit around the Earth, we have apogee and perigee. Similarly we have for other objects in orbits around the Sun – aphelion, perihelion; the Moon – apocynthion, pericynthion; Jupiter – apojove, perijove; and a star – apastron, periastron, etc.

Synodic Period

The time taken for a planet to go around its orbit once, the orbital period, is normally called the sidereal period of the planet. It is the fundamental quantity required in determining the orbit of the planet. However, our observations of the planets are made from the moving Earth, and so we do not observe the sidereal period directly, but the period required for the planet to return to the same relative position with

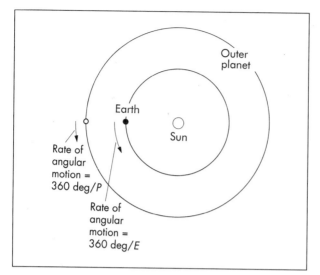

Figure 5.21.
Relationship between synodic and sidereal periods.

respect to the Earth and Sun. The interval between successive returns to the same relative position (e.g. between two oppositions) is called the synodic period of a planet. Since we are observing from the Earth, it is the synodic period that we are able to measure, and the sidereal period must then be derived from it. From Figure 5.21, we may find the sidereal period:

$$\text{Relative angular velocity} = \frac{360}{E} - \frac{360}{P} \quad (5.11)$$

$$= \frac{360}{S} \quad (5.12)$$

where E is the sidereal period of the Earth (1 year), P is the sidereal period of the planet, S is the synodic period of the planet which gives, for an outer planet, working in units of years (the sidereal period of the Earth, so that $E = 1$),

$$P = \frac{S}{S-1} \quad (5.13)$$

and we may similarly find for an inner planet

$$P = \frac{S}{S+1}. \quad (5.14)$$

Exercises

5.1 What is the azimuth of the Sun on rising at the summer solstice as seen from the old Greenwich Observatory? (Greenwich Observatory: latitude 51° 29′; longitude 0°.)

5.2 From above what latitude is Betelgeuse a circumpolar star (ignore precession)? (Betelgeuse: $RA_{(2000)}$ 5 h 55 m 10 s; $Dec_{(2000)}$ +7° 24′ 26″.)

5.3 Calculate the position of Sirius on 5 November 1992, by allowing for the effects of precession. (Sirius: $RA_{(1900)}$ 6 h 41 m; $Dec_{(1900)}$ −16° 35′; obliquity of the ecliptic, $\varepsilon = 23° 27.$)

5.4 Calculate the synodic periods of
(a) Venus
(b) Jupiter
(c) Pluto
(Sidereal orbital periods of Venus 0.615 yrs; Jupiter 11.862 yrs; Pluto 247.7 yrs.)

5.5 Can
(a) a lunar eclipse occur at new Moon?
(b) a solar eclipse occur at the summer solstice?

Introduction

Telescope mountings divide into two distinct sections. The first is the set of mechanical components that hold the optics of the telescope in their correct relative positions, and allow them to be collimated and to be focused. This section is normally called the telescope tube, and reference is made to it in Chapters 1, 2, 3 and 8. Here, therefore, we are concerned with the second section, which is the set of mechanical components that enables the telescope tube and the optics to be pointed at the object to be studied, and in most cases then to compensate for the Earth's rotation (i.e. to track the object) automatically. Reference to this section of the mounting has also been made elsewhere (Chapters 1, 2, 3, 4, 5 and 8), but we now look at its specifications in more detail. In this chapter, we shall use the term *mounting* to refer only to this second section from now onwards.

The basic requirements for the mounting are:

1 to enable the object to be observed to be found in the sky as quickly and easily as possible;

2 to enable the object to be held steadily in the field of view as eyepieces are focused or changed, and cameras, photometers, spectroscopes, etc. are attached to the telescope tube;

3 to enable the telescope to follow the object as it moves across the sky, because of the Earth's rotation, for periods of time up to several hours; and

4 to provide the observer with convenient access to the eyepiece or ancillary instrumentation.

All these objectives are achievable to almost any desired level of performance, but as ever, the cost of the mounting will rise steeply as the specifications on its performance are increased. For a small telescope, it is generally necessary to compromise at some point, or the cost of the mounting may become many times the cost of the telescope. With major telescopes, initial acquisition may be to an accuracy of a few seconds of arc, and tracking to better than a fraction of a second of arc over an hour. There are innumerable designs for telescope mountings, but they fall into the two categories of equatorial (Figure 2.18) or alt-az (Figure 2.19), and we shall consider the equatorial first.

Equatorial Mountings

The objectives for a telescope mounting are most straightforward to achieve with designs of this type. The telescope tube can be moved about two perpendicular axes, one of which is parallel to the Earth's rotational axis. Movement of the telescope around this axis (the polar axis) therefore changes only the hour angle or right ascension (Chapter 4) that is observed, while movement around the other axis (the declination axis) changes only the declination that is observed. The position of the telescope in RA (or HA) and Dec can easily be determined from suitably graduated 360° protractors mounted on the two axes, and known as setting circles. The Earth's rotation can be compensated by a motor that drives the telescope around the polar axis, from east to west at a constant rate of one revolution in 23 hours 56 minutes (the sidereal day – see Chapter 4). The main variations on the basic design are shown in Figure 6.1. Any of these designs, or their many variants, can make a suitable mounting for a small telescope. However, a prime requirement of any mounting is for stability against vibrations, and the designs with single supports and unbalanced weights such as the German and fork mountings, will, all other things being equal, be poorer in this respect than the English and modified English designs. In selecting a mounting, if there are alternatives, then in general the thicker the axles, the larger the bearings and the more massive the construction, the better.

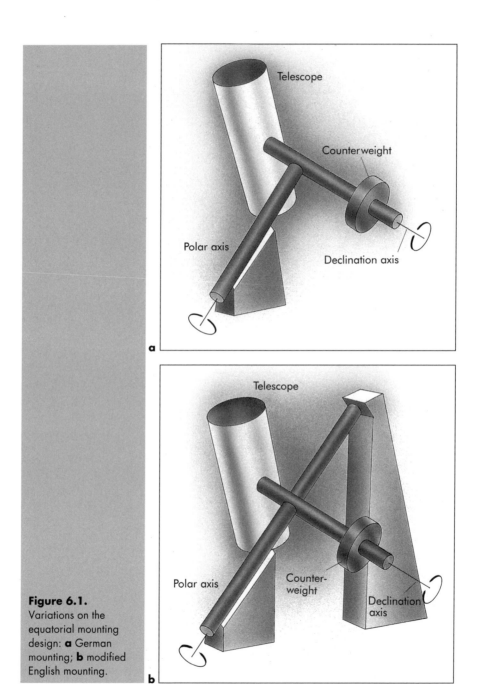

Telescope

Counterweight

Polar axis

Declination axis

a

Telescope

Polar axis

Counter-
weight

Declination
axis

b

Figure 6.1.
Variations on the
equatorial mounting
design: **a** German
mounting; **b** modified
English mounting.

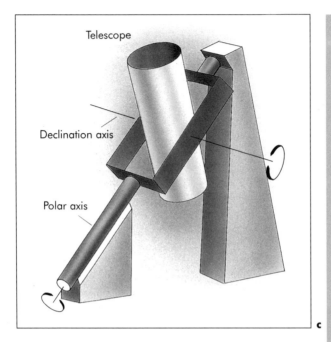

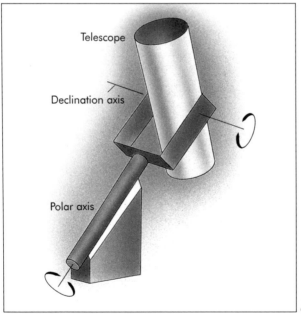

Figure 6.1. *(cont'd)* Variations on the equatorial mounting design: **c** English or yoke mounting; **d** fork mounting.

Alt-Az Mountings

Alt-az mountings also enable the telescope to be moved around two perpendicular axes, but this time one of the axes is vertical and the other horizontal (Figure 2.19). There is thus no simple way of relating the position of the telescope with respect to its mounting to the point in the sky that is being observed. There is no simple means of enabling the telescope to track an object either, since this will require motions in both axes, and at non-uniform rates. The alt-az mounting, however, does have one big advantage – it is much easier to make it stable and robust than any of the equatorial designs, because the weight, and therefore the stresses, always act in the same directions. In other words, a given level of performance in terms of stability can be achieved at lower cost with an alt-az than with an equatorial mounting. This has resulted in many of the major telescopes commissioned recently and some smaller commercially produced telescopes to have alt-az mountings. The extra cost of the motors for both axes and the computers to control them is more than made up by the reduction in the overall cost of the mounting for such instruments. Most alt-az designs are like the fork mounting (Figure 6.1d) with the polar axis pointing to the zenith, and so there is little to say about them. One particular variant, however, has become popular recently because of its cheapness and portability, and that is the Dobsonian design (Figure 6.2). It uses Teflon strips as the bearing material, giving very smooth adjustment of the telescope's position. The whole mounting can be placed on a low platform that, by rotation over an inclined plane, will give tracking of objects in the sky over short intervals of time, so overcoming the main drawback of the alt-az. There remains one problem with the alt-az mounting, however: if an object is followed as it moves across the sky, the image plane rotates with respect to the telescope. Large instruments use image rotators to counteract this effect, but there is little that can be done for smaller instruments.

Making Your Own Mounting

Just as it is possible to make your own optics for a telescope to a standard as good as or better than many

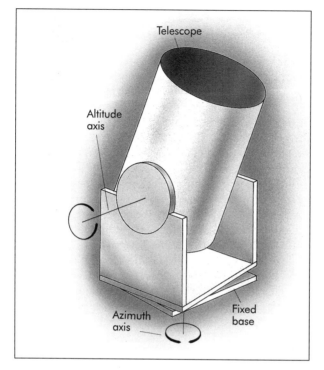

Figure 6.2.
Dobsonian mounting.

commercially produced items (Chapter 3), so it is possible for a competent DIY enthusiast to produce a more than adequate mounting. There is no standard approach, however, to producing a mounting comparable with the grinding and polishing of the optics, which varies little in its essentials from one telescope maker to another. The intending mounting constructor should take to heart the requirements for a mounting (described above), see as many different designs as possible, perhaps by joining a local astronomical society, consult the frequently published articles in the popular astronomy journals on the subject, and then finally produce a design suited to his or her needs, abilities, available equipment and materials.

Alignment

An equatorial mounting needs to be aligned so that its polar axis is precisely parallel to the Earth's rotational axis if it is to track objects well, and if its setting circles are to be accurate. For a permanently installed telescope, this alignment only has to be done once, and so the observer can afford to spend sufficient time to get it done well. For a portable telescope, the alignment will have to be done each time the telescope is set up, and will therefore normally be accomplished to much lower accuracy.

If the telescope has a cross-wire eyepiece and setting circles, the initial setting-up can be done by setting the telescope to the coordinates of the pole star, then finding the pole star and centring it on the cross-wires by moving the polar axis in azimuth and altitude. Even without setting circles, aligning the telescope by eye to be parallel to the polar axis and then finding Polaris will set the mounting up to within a degree or two. Some commercially produced telescopes are provided with a rifle sighting scope which may be attached to the mounting in order to sight on the pole star in a similar way. An arrangement of this sort is almost essential for a portable telescope if considerable amounts of time are not to be wasted every time it is used.

A permanently installed telescope mounting can be aligned much more precisely after this initial setting up, using observational tests. Corrections to the altitude of the polar axis are found by observing a star between 30° and 60° declination, about 6 hours east or west of the observer's meridian. The star should be centred on the cross-wires of a high-power eyepiece, and tracked using the telescope drive. If, after a while, a star to the east of the meridian is observed to drift northwards in the eyepiece, then the altitude of the polar axis is too high. A star to the west of the meridian would drift to the south in the same circumstances. If the easterly star drifts south, or the westerly star north, then the polar axis is set to too low an altitude. After a suitable adjustment to the axis, the procedure is repeated until any remaining drift is acceptable. Alignment will clearly have to be done much more precisely if the observer intends, say, to undertake photography with exposures of several hours rather than visual work. A similar procedure may be used to align the polar axis in azimuth. Two stars are selected which are 10 or 20 degrees north and south of the equator, and differing in right ascension by a few minutes of arc (reference to a reasonable star catalogue will probably be necessary for this – Appendix 2). When the stars are within an hour or so of transiting the observer's meridian, with the drive *off*, the telescope is set just ahead of the leading star (the one with the smaller right ascension). A stopwatch is started as that star transits the vertical cross-wire. The telescope is then moved to the declination of the second star, and the time interval until that star transits is measured. The correct time interval between the two transits can be found from the difference in their right ascensions. If the measured time interval is too small, then the polar axis is aligned to the east; if the time interval is too large, then the polar axis is to the west of its true direction. Adjustments to the position of the polar axis are made and the procedure again repeated until satisfactory.

Setting Circles

With a well-aligned equatorial mounting, 360° protractors suitably graduated, and called setting circles, can easily be mounted to allow the position of the telescope in the sky in terms of right ascension or hour angle and declination to be measured; these are then a great aid to finding objects in the sky. The declination can be read simply from a protractor graduated from 0° to ±90° attached to the telescope and with a pointer fixed to the mounting (or vice versa – see Figure 6.3). Even without RA or HA setting circles, the declination setting circle is very useful. The telescope can be set accurately to the declination of the required object, and pointed roughly in the right direction in RA by looking at the constellations. The telescope can then be swept

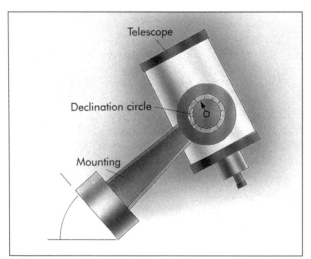

Figure 6.3.
Declination circle.

Figure 6.4. RA circle.

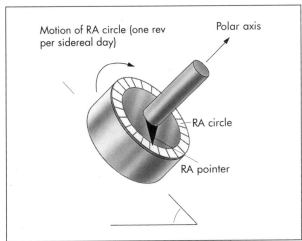

Motion of RA circle (one rev per sidereal day)

Polar axis

RA circle

RA pointer

in RA, without changing its declination, and the required object will often be found rapidly.

Hour angle is equally easily determined using a protractor graduated from 0 h to 24 h and attached to the polar axis of the mounting. Unfortunately, the hour angles of objects vary with time and longitude, and the correct setting has to be calculated (Equations 4.1, 4.2 and 4.12) every time it is wanted from the local sidereal time. A setting circle reading RA directly is therefore much more useful. This also is a protractor graduated from 0 h to 24 h, but driven at sidereal rate to allow for the motion of the sky (Figure 6.4). It will normally need to be set at the start of an observing session. Its correct setting can be calculated from first principles, but usually it is simpler to find a bright object with a known right ascension in the telescope, and then to set the RA circle to read that right ascension correctly. Thereafter, the circle will read RA directly, unless its drive is turned off at any time. Some modern small mountings use digital read-outs for the telescope's position, but the principle does not differ from the use of the circles.

Guiding

During long exposures, it will be found with most mountings and drives that the object being observed wanders about in the field of view. This may be due to inadequate alignment of the mounting (see above), to changing flexure in the mounting, or to non-uniform or incorrect drive speeds. Even a perfect mounting and drive would not track an object precisely over several hours however, because refraction in the Earth's atmosphere changes the apparent positions of objects in the sky (Equation 5.7). For visual work, such inaccuracies in the tracking are unimportant unless their magnitude is huge; but if images are being obtained by photography or CCD detector, or the light is being fed into a photometer or spectroscope, then the inaccuracies have to be corrected or the images will be trailed, or the light lost from the instrument. Such correction of the drive is known as guiding.

Conventionally, guiding is undertaken by the observer who looks at the object through a second telescope attached to the main telescope and called the guide telescope. With a cross-wire eyepiece in the guide telescope, the observer manipulates the slow-motion adjustments of the mounting to keep the object centred on the cross-wires throughout the exposure. This procedure requires considerable practice for success, and can become tedious during long exposures. It also requires the guide telescope to be not much smaller than the main telescope, or the fainter objects will not be visible. For these reasons, many observers use autoguiders of various designs. An autoguider is a device that can detect the object being observed, and then generate an error signal if the object wanders away from its correct position. The error signal is used to operate the slow motions in RA and Dec in such a manner that the object is brought back to its correct position. The autoguider may be attached to the guide telescope, or the field of view of the main telescope may be sampled, and a nearby object to the one of interest used for guiding. This latter process of off-axis guiding is also often useful for fainter objects generally, when a nearby brighter object can be selected for easier guiding. Autoguiders are available commercially, and are advertised in the popular astronomy magazines (Appendix 2). With CCDs, another approach is possible, and is based upon the ease with which many CCD images may be added together. An exposure is chosen which is sufficiently short, so that *unguided,* the images show no trace of a trail. Multiple images are then obtained with this basic exposure, aligned with each other and added together to give a single image with the required total exposure.

Modern Commercial Mountings

Commercially produced telescopes (Chapter 3) are invariably provided with mountings. Though the optics of such instruments are generally excellent, they often leave something to be desired in terms of the quality of their mounting. At the worst there are the "single pivot", "ball and claw" and related types of mounting (Figure 6.5) which should be avoided completely. Fortunately this type of mounting is now less commonly found than in earlier times. Variations on the mounting designs discussed above are to be found for most commercial telescopes, with the German and fork designs probably being most popular. Unfortunately, they are often too lightweight for the size of telescope they are required to carry. This does not mean that they are likely to break, but that vibrations will be only slowly damped out, thus limiting their usefulness for photography or CCD imaging. Mountings have generally become heavier and more stable in the last few years, and in some cases such as the Celestron and Astro-physics ranges, heavy-duty German mounts are now available as options that will provide a reasonable level of rigidity. Nonetheless, the prospective purchaser of a telescope should still pay close attention to the mounting, and make use of any opportunity to try it out in practice.

Figure 6.5. Single pivot mounting (to be avoided!).

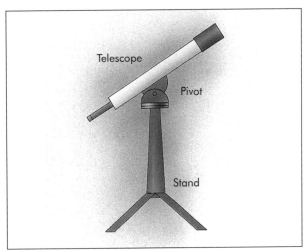

Many modern mountings are now linked to a small computer, which can provide read-outs of the RA and Dec of the telescope. Usually the positions of hundreds or even thousands of interesting objects in the sky are also stored on the computer, and can therefore be found rapidly in the telescope. Slight changes to the drive speed from the sidereal rate to match the motions of the Sun and Moon are also often available. In at least one case, the computer can be used to compensate for inaccurate alignment. Two stars are found in the telescope, and their details fed into the computer that thereafter will correct the displayed positions to correspond to the actual point in the sky that the telescope is observing.

Section 3

Observing

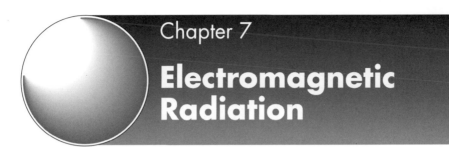

Chapter 7

Electromagnetic Radiation

Introduction

Almost all our information in astronomy is obtained by the electromagnetic radiation travelling from the object to the observer. Apart from those few objects within the solar system that we have been able to investigate directly, cosmic rays, neutrinos and, in due course, gravity waves are the only other information carriers likely to tell us about the universe as a whole. The electromagnetic radiation wave consists of a magnetic wave and an electric wave whose directions are orthogonal, and which vary sinusoidally. The frequency of the sinusoidal variation is called the frequency of the wave and denoted usually by ν. The separation of successive crests or troughs gives the wavelength, λ. The product of ν and λ gives the wave's velocity. In a vacuum this has a constant value, denoted by c, of 299 792 500 m s^{-1}.

$$\lambda \nu = c. \qquad (7.1)$$

In other media the velocity is reduced from c by an amount given by the refractive index, μ, of the material:

$$v = \frac{c}{\mu}. \qquad (7.2)$$

Since even materials normally regarded as opaque have a less than infinite absorption, light penetrates them to a certain extent and so they have a refractive index. We can thus talk, for example, about the velocity of light in a brick!

In a medium where the electromagnetic radiation has a velocity less than c, the frequency remains constant, so from Equation (7.1), with v replacing c, the wavelength becomes shorter. As is well known, the special theory of relativity postulates that the velocity of electromagnetic radiation in a vacuum is the maximum possible for any normal particle, and experiments have confirmed this many times. However, because of the reduction in velocity given by Equation (7.2), in any medium *other* than a vacuum it is possible for, say, a particle to exceed the local velocity of light. When the particle is charged, as for example is the case with most cosmic ray particles, a type of radiation is produced known as Čerenkov radiation. This is the electromagnetic equivalent of the sonic boom of a supersonic aircraft. It is of significance in that it results in noise spikes in photomultipliers and CCD detectors (see Chapter 10).

Intensity

The equation for the electric component of an electromagnetic wave is

$$E(x,t) = E_0 \sin\left[\frac{2\pi x}{\lambda} + 2\pi t v + \phi\right] \qquad (7.3)$$

where $E(x,t)$ is the magnitude of the electric vector at position x and time t; E_0 is the amplitude of the wave and ϕ is the phase at $t = 0$, $x = 0$.

The intensity of the radiation is given by E_0^2. It has units of energy per unit area and per unit wavelength or frequency interval, and for convenience various different units are used over the spectrum. Thus, in the radio region, the units are janskys (1 Jy = 10^{-26} W m^{-2} Hz^{-1}) at shorter wavelengths, the units are W m^{-2} Hz^{-1}; while at the shortest wavelengths, electron volts (1 eV = 1.6022×10^{-19} J, the energy an electron gains in falling through one volt) are often used for the energy of one photon. Determination of the intensity of the electromagnetic radiation from a source is the most fundamental measurement made in astronomy. Even for visual work, although we do not normally express it in this fashion, an image is registered because the eye determines different intensities falling on to different parts of the retina.

Photons

Although for many astronomical purposes we can regard electromagnetic radiation as a wave, it actually has a dual nature, and sometimes behaves as a particle, known as a photon or quantum of light. The converse is also true: electrons, which we usually regard as particles, can sometimes behave as waves, allowing among other things the production of electron microscopes. For astronomy at visual wavelengths, the particle nature of light is mostly of significance in the way most types of detectors work. Thus the eye, photographic emulsion, CCD and photomultiplier all detect photons rather than waves. Quantum mechanics is required to deal with the complete situation; here we just note that when behaving as a particle, the photon of radiation has an energy, E, given by

$$E = h\nu = \frac{hc}{\lambda} \qquad (7.4)$$

where h is Planck's constant (6.6262×10^{-34} J s).

The units for photon energy are joules but, as noted above, electron volts are often used because they lead to more convenient numbers. Thus the visual spectrum ranges from 1.8 eV (red) to 3 eV (blue).

Polarisation

A second fundamental property of radiation, but one less commonly studied in astronomy, is the state of polarisation of the radiation. If the direction of the electric vector (or magnetic vector) in space in a beam of radiation changes randomly on the time scale of $1/\nu$, then the radiation is unpolarised. If the direction of the electric vector is constant, then the radiation is plane polarised. If the direction rotates at the frequency of the radiation, it is circularly polarised (clockwise or counter-clockwise, according to the direction of rotation of the vector). If the amplitude also varies at the frequency of the radiation, then we have elliptically polarised radiation (the tip of the electric vector would trace out an ellipse if you were to look down the beam). Most radiation from astronomical sources is unpolarised, or nearly so. Some sources, however, may be highly polarised (for example, the Crab nebula, and active galaxies), and much information can then be gleaned from the details of the polarisation.

Range

Electromagnetic radiation extends, in theory, from an infinite wavelength ($v = 0$, $E = 0$), to a zero wavelength ($v = \infty$, $E = \infty$). In practice, we can detect from wavelengths of a few tens of kilometres to 10^{-14} m or so. By tradition and for convenience, different parts of the spectrum are labelled by different names (Figure 7.1), although the nature of the radiation is the same at all wavelengths. The nature of the interaction of radiation with matter, however, does change in different parts of the spectrum, and this results in the differing detectors and telescopes used by radio, optical and X-ray astronomers. It also means that we can obtain different types of information on the universe from different parts of the spectrum. Thus, as a generalisation, at long wavelengths, we have direct induction of electric currents, free–free and synchrotron radiation; in the microwave and infrared regions, the interaction is with rotating and vibrating molecules; in the optical region, it is with the outer electrons of atoms and molecules; in the ultraviolet and X-ray region, with inner electrons and ionisations; and at the shortest wavelengths, directly with nuclei. Thermal emission can occur at all wavelengths but is more generally important at longer wavelengths.

Astronomy started off with visual observations because that is the spectral region to which our eyes are sensitive, and because the atmosphere is reasonably transparent in that region. Observations of the Sun in the infrared and ultraviolet regions were being made by the early nineteenth century, and of the Moon and other objects 50 years later, again because the atmosphere is reasonably transparent in those regions. The radio region was developed from the 1940s, yet again because the atmosphere is reasonably transparent in that region. All other spectral regions had to await balloons, rockets and spacecraft from the 1960s onwards for their exploitation because there the atmosphere is opaque (Figure 7.2).

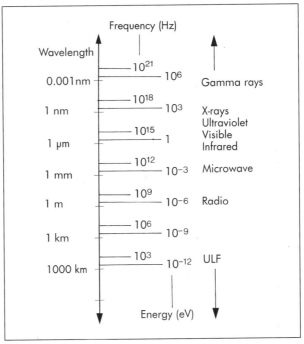

Figure 7.1. Range of electromagnetic radiation.

Measurements

Figure 7.2. Atmospheric transparency.

From Equation (7.3) we may see that the only parameters of electromagnetic radiation that we may measure are its amplitude (E_o), wavelength or frequency, phase, plus its direction and state of polarisation. Only in the

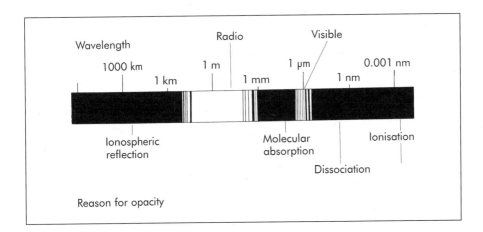

radio region are all these quantities routinely deter-
minable. At all shorter wavelengths, the measurements
are generally of intensity (E_0^2), and wavelength, plus
occasionally linear polarisation. Observations are
divided on a practical basis, however, into several dis-
tinctive areas.

Photometry

This is theoretically the measurement of intensity at a
single wavelength. In practice, it is the intensity integrated over the pass-band of the filter being used.
Separate photometric measurements may be made at
several different wavelengths, which in the limit as the
number increases and pass-bands decrease approaches
spectroscopy. The details of photometry are covered in
Chapter 11. Numerous photometric measurements
over a two-dimensional field (i.e. measuring direction
as well) give us an image of that field.

Spectroscopy

This is the measurement of intensity as a function of
wavelength; as previously mentioned, it is the limiting
case for photometry over numerous pass-bands.
Occasionally the spectrum may be obtained simultaneously at a number of points over the field, usually
leading to a transect of the object in the field of view
(long slit spectroscopy). It is considered in more detail
in Chapter 12.

Polarimetry

This is the measurement of the degree and direction of
linear polarisation, or, much more rarely, of the state
of elliptical polarisation. The measurements are often
extended over two dimensions to give a polarisation
map of an object. It is a highly specialised technique
and is not considered further in this book.

Chapter 8

Visual Observing

Introduction

In this chapter it is assumed that the reader has access to a small to medium-sized telescope (10–30 cm, or 4–12 inches) on an equatorial mount, with a motor drive, and intends to use it for visual work. If the telescope is not on an equatorial mounting, then most of the observations will still be possible but will generally be more difficult. If the telescope is smaller than 10 cm, then many of the fainter objects will be difficult or impossible to see. If the telescope is larger than 30 cm – congratulations! Other types of observing such as imaging with photographic emulsion or CCD, photometry and spectroscopy are considered in later chapters. Visual work is an aspect of imaging (Chapter 9); nowadays it is sometimes regarded as inferior to methods providing a permanent record – it certainly can be a subjective process (see later), and is limited in the wavelengths and intensities detectable. Nonetheless, with care and practice it can still give very valuable results; after all, for half the time since their invention, telescopes could be used with no other type of detector than the eye. Furthermore, one of the principal joys of astronomy is seeing for yourself the mysteries and wonders of the universe, and it should never be forgotten that one of the great strengths of the science is that the vast majority of astronomers study the subject for pleasure and interest (how many amateur solid-state physicists do you know?).

General and Practical Considerations and Safety

Whether in an observatory or not, telescopes are often used in damp conditions, and obviously at low levels of illumination. It is therefore advisable for all equipment to be of low voltage. Any high voltages should always be taken through rapid circuit breakers to avoid the possibility of electrocution. Other safety considerations are to wear appropriate clothing, never to rush around the site, and if not working with someone else (see below) to let a responsible person know when and where you will be observing and when you expect to finish. That person should have instructions to take appropriate action (ranging from calling out "Are you all right, dear?" from the back door to calling the emergency services) if you do not return on time. Heavy items, such as counter-weights, should be handled carefully and attached securely to the telescope.

Certain factors apply to all types of observations. One of the most important of these is the selection of observing site. This has already been discussed in Chapter 2. Here, there is just the reminder that if there is any choice in the matter, the site should be as well away from artificial light sources as possible. In choosing a site for, say, a small portable telescope, the question of personal safety should not be neglected. However, since being away from artificial light will usually mean that the site is isolated. It is a sensible precaution therefore to take a companion in such circumstances, and always to obtain prior permission if using private ground.

If the telescope is not in some sort of observatory, then movable screens to protect it from the wind and perhaps to obscure lights will be found very useful. A commonly encountered problem is that of misting of the optics, or dewing up as it is known. If observations seem to be becoming more difficult, always check to see if dew is forming, since even a very thin layer will ruin the image. The dew should *never* be wiped off the optics, since coatings can be damaged and surfaces scratched. In the cases of refractors and Schmidt–Cassegrain telescopes, a long extension of the telescope tube beyond the objective, known as a dew cap, can reduce the formation of dew. It is also possible to add a low voltage heating cable around the objective, but if this is too powerful it may induce convection currents and cause the images to deteriorate. If these precautions fail, then a low voltage blow hair drier, of the type sold for campers and caravaners, can be used to evaporate the dew. While this also runs the risk of inducing convection currents, they are likely to die away shortly after the drier has ceased to be used. In heavy dewing conditions, it may be necessary to clear the optics at intervals as short as a few minutes.

Even on clear nights, images will often be poorer than the diffraction limit of the telescope. This is due to turbulence in the atmosphere causing the image to scintillate, an effect also known as twinkling. The magnitude of the effect is known as the seeing. From the very best observing sites under the very best conditions, it can mean that a stellar image is about 0.25" across. The best that can be expected from an average site is about 1", and it will more typically be 2" and on poor nights as much as 5".

As discussed in Chapter 2, the choice of eyepiece is important. As a general rule, the lowest power that will reveal the objects of interest should be used, and the maximum usable magnification will reduce sharply as the seeing deteriorates. Except under the very best conditions, high powers will give fuzzy images that lack detail. Eyepieces can also dew up, and the observer should be careful to avoid breathing on them. The problem can be reduced if eyepieces not in use are kept in individual lint-free cloth bags in an inside pocket, so that they are warmer than their surroundings when brought out for use.

Finding

Before anything can be observed, the telescope must be pointed in the right direction – that is, the object must be found. If the mounting has setting circles (Chapter 6), and the right ascension (or hour angle) and declination (Chapter 4) of the object are known, then the telescope can be set directly on to the object. Some telescopes are linked to computers that store the positions of numerous objects and can therefore be set very quickly. If such a device is not available, then the positions of the stars, galaxies, etc. can be looked up in catalogues or read off star charts (Appendix 2). A planisphere is useful for ascertaining quickly which constellations are in the sky at a particular time of the night throughout the year.[4]

The positions of the Sun, Moon and planets can be found for the date of observation from the *Astronomical Almanac,* or from the popular astronomical journals (Appendix 2). The positions of ephemeral objects like comets and novae are sometimes listed in the popular journals or even in newspapers. The national astronomy societies and the IAU (Appendix 3) may also send out details, though sometimes this requires an additional subscription. The observing details for a new object like a comet will also often be found quickly by a web search. If the mounting has been set up correctly and the setting circles are accurate, then the object should be in the field of view. However, most setting circles on small telescopes can only be set to a quarter of a degree or so. The eyepiece that gives the widest field of view with the telescope, normally the lowest power eyepiece available (Equation 2.8), should therefore always be used for finding. If the main instrument has a finder telescope, then this will normally have an even larger field of view, and should be used before trying the main telescope. If the object is not in the field of view initially, then it will usually be possible to find it by nudging the telescope by small amounts to search around the area. If it still does not come into view, then check for something being amiss. Common problems include:

- forgetting to take off a mirror cover;
- one or more of the optical components being covered in dew;
- misreading the setting circles;
- looking up the details of the wrong object in the catalogue;
- miscalculating the sidereal time (do not forget to take off any summer time corrections to the local civil time);

---

[4] This is a device that shows the stars visible from a particular latitude at any time of the night or year. It comprises two circular sheets pivoted at their centres. The bottom sheet has a map of all parts of the sky visible from the latitude. The top sheet has an elliptical window that represents the hemisphere of the sky visible at a particular instant. Scales on the edges of the sheets allow the top one to be rotated to show the stars visible at a particular instant. Planispheres may be purchased quite cheaply from a good book store.

- not correcting or correcting the wrong way for precession if the data are from an old source;
- misalignment of the mounting.

Once the object has been found, then it may be centred in the eyepiece by adjusting the position of the telescope, and higher-power eyepieces substituted if required.

If setting circles or the accurate position of the object are not available, then it must be found by a search. This is not difficult for any object bright enough to be seen with the naked eye. The telescope is roughly aligned towards the object by sighting along its tube. The object should then be visible in the finder, or if not, findable by a search around the area while looking through the finder. Centring the object in the finder should then enable it to be found in the main telescope. For objects too faint to be seen with the naked eye, but bright enough to be seen through the finder telescope, the procedure is similar, but the initial setting will have to be done by estimating where to point the telescope by comparison with a nearby object bright enough to see. For objects visible only in the main telescope, a finder chart needs to be drawn from a star chart or catalogue. Sometimes ready-prepared charts will be published in the popular astronomy journals for comets etc. There are also several books that contain finder charts for the Messier objects, and some other nebulae (Appendix 2). The finder chart should show the pattern of stars around a nearby bright object, especially those along the line to the object of interest (Figure 8.1). The bright object is then found, and the star pattern identified. Remember most telescopes give an inverted image, so for this purpose it is useful to draw the finder

Figure 8.1.
Schematic finder chart.

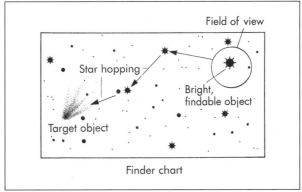

Finder chart

chart on tracing paper so that it can be viewed from either side. The telescope is then moved along the star pattern until the desired object comes into view, a process sometimes called star hopping. For this sort of procedure it is useful to know the size of the actual field of view of the eyepiece in the sky so that the finder chart is of the appropriate scale. This can be calculated (Equation 2.8) or measured. To measure the field of view, set the telescope on to any object whose declination is known, position that object to one side of the eyepiece, and turn off the telescope drive. The time taken for the object to drift centrally across the eyepiece should be measured (if the object is set to the wrong side of the eyepiece, so that it drifts out of the field of view, just bring it back and set it to the other side). The angular size of the field of view is then given by

$$\text{Field of view} = 4.2 \times 10^{-3} t \cos \delta \quad \text{degrees} \quad (8.1)$$

where t is the time, in seconds, for the object to drift across the field of view.

It is helpful to note prominent asterisms (easily recognisable groups of stars) on the star chart and move from one such to the next to ensure positive identification at each stage of the route, or it is easy to get confused and lose the way.

Moon

Though familiar to everyone and sufficiently close and bright for some details to be seen with the naked eye, the Moon remains one of the most rewarding objects for telescopic observation. It is ideal as a first object to look at, being easy to find, and with a wealth of features discernible by even the most inexperienced observer. It is also an object that may be observed from the worst imaginable observing sites, even through an open window from an apartment in the centre of a brightly lit city, provided only that the window faces in the right direction. Yet at the same time, it is an object worthy of continuing and detailed study. For despite the various spacecraft missions to and around the Moon, there remain puzzles about it that deserve attention.

All the features visible on the Moon from the Earth have long ago been mapped and named, and as an initial task the observer should become familiar with a few of the more prominent ones. These would be the lunar maria and some of the larger craters, and they will then provide reference points for finding more obscure features. For this purpose a map of the Moon will be needed. However, care needs to be taken in the choice of such a map for two reasons. Firstly, in most astronomical telescopes the image is inverted (Chapter 2). The Moon is therefore seen with its south pole at the top of the field of view. Most maps of the Moon published in astronomical sources show the Moon in this orientation, and so are easy to correlate with what is seen in the telescope. Other sources, however, especially those with space-related interests, plot maps of the Moon as it would be seen with the naked eye, with the north pole at the top. Such maps are very confusing as references for observing and should be avoided. The second problem is that the appearance of the lunar surface changes very considerably with the phase. Thus near full Moon, many of even the largest craters are almost invisible. Small craters may only be detectable for one or two days at exactly the right phase when their shadows make them prominent. For this reason, some lunar maps show features over the whole surface with a constant angle of illumination (i.e. with an identical altitude for the Sun as seen from each point on the Moon). They are therefore quite misleading as a guide to the appearance of the whole Moon, though useful for more detailed studies.

Craters are by no means the only type of object visible on the Moon; there are the various other features, known as rilles, domes, scarps, valleys, mountains, lava flows, crater rays, etc. To become reasonably familiar with the Moon is therefore a task likely to take a year (or a lifetime) of dedicated observing. During such a programme it is of interest to determine the smallest objects that can be discerned. Under the best observing conditions, these are likely to be about one kilometre or so in size. But, as already mentioned, small features will often only be detectable for a short time while the shadows are exactly right, and since such a time must also coincide with very good observing conditions, many small objects may only actually be visible on one or two occasions throughout the whole year. It can also be of interest to try and detect the new Moon as early as possible, or the old Moon as late as possible. It might seem that the Moon should be detectable right up to the instant of new Moon, but because it is then close to the Sun in the sky, it is swamped by the scattered solar light. Detections within 30 hours either side of new Moon can generally be regarded as an above average achievement.

Lunar eclipses (Chapter 5) do not have the same significance as solar eclipses, but can nonetheless be interesting to observe. The Moon is, of course, full at the time of an eclipse, and so little detail will be visible. The main interest is in the details of the Earth's shadow. Depending upon atmospheric conditions, and events such as major volcanic eruptions, this can vary from almost black, so that the Moon disappears, to quite a bright reddish hue. The colour in the latter case is due to light being scattered in the Earth's atmosphere towards the Moon. An apparently somewhat similar phenomenon is "the new Moon in the old Moon's arms". When the Moon is a narrow crescent, the dark portion can sometimes still be seen. This, however, is due to reflected light from the illuminated part of the Earth reaching the Moon (Figure 8.2). If you were actually on the "dark" part of the Moon at such a time, then since the Earth is over five times as reflective as the Moon, and three and a half times larger, the amount of light received would be about seventy times as much as we receive from the full Moon – more than bright enough to work, read and move around the surface.

There are several aspects to lunar observing suitable for more advanced work. As mentioned in Chapter 5, though the Moon nominally keeps the same portion of its surface towards the Earth, its speed around its orbit varies owing to the ellipticity of the orbit, and so sometimes its rotation gains or loses slightly on its orbital motion. We can thus see about 59% of the total surface

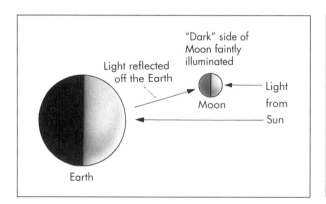

Figure 8.2. The new Moon in the old Moon's arms.

area of the Moon. This slight wobble is known as libration, and observing the limbs of the Moon at times of maximum libration enables glimpses to be had of what is normally the far side. The heights and depths of features on the Moon can be measured relative to their immediate surroundings quite simply. Ideally a micrometer eyepiece (Chapter 2) is used to measure the angular size of the shadow of the object, and its distance from the lunar limb. However, in the absence of such an eyepiece, an ordinary cross-wire eyepiece, normally used for guiding, can be used instead. The angles are then measured by drifting the image with the telescope drive turned off, timing how long it takes to pass over the cross-wire, and converting the time into an angle using Equation (8.1), or more conveniently in seconds of arc:

$$\text{Angular distance} = 15t \cos \delta \quad \text{seconds of arc.} \quad (8.2)$$

The height of the feature in metres is then given in terms of the angles shown in Figure 8.3 by

$$\text{Height} = 1900L \left[\frac{R-D}{R} - \frac{(R-P)\sqrt{(2DR-D^2)}}{R\sqrt{(2PR-P^2)}} \right] \text{metres}$$

$$(8.3)$$

where L, D, P and R are in units of seconds of arc.

One controversial area in which observations can be valuable is that of transient lunar phenomena, or TLPs.

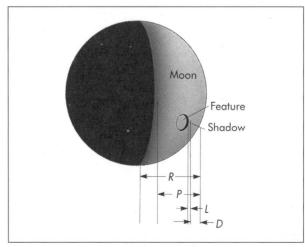

Figure 8.3. Finding the height of a feature on the Moon.

There have been frequent claims of observations of short-term changes to features on the Moon, but many astronomers remain sceptical. The typical TLP consists of an apparent veiling, as though by a thin mist or dust cloud, of a small area on the Moon. The observation is generally therefore that the feature is less clearly defined than usual rather than a specific alteration, which accounts for some of the scepticism. However, on a few occasions, the obscuration has been much more pronounced, and light emissions have sometimes been claimed. The events cluster in a few areas such as around the central mountains of the craters Aristarchus and Alphonsus. They are perhaps more common when the Moon is near perigee, which could be due to the increased tidal stresses at that time. If they occur, then they may be due to trapped gases making their way to the surface and producing a short-lived dust cloud at the surface from dust entrained along the way. If the light emissions are real, then these could be glow discharges produced by the build-up of static electrical potentials (i.e. the lunar equivalent of lightning). The whole subject is a fascinating one, and an area where observations made with small telescopes can potentially make a real contribution to science.

Occultations (Chapter 5) of stars by the Moon are frequent because of its large angular size and rapid motion across the sky. They are studied using high-speed photometers (Chapter 11) as a means of determining a star's angular size, and to obtain details of close double stars. However, the rapid flickering upon which such measurements are based will generally not be detectable by the eye. On some occasions, though, especially if the relative motion is close to parallel to the limb of the Moon, it may be possible to see the double dimming as a double star goes behind the Moon, and very occasionally the star may reappear briefly when seen through a valley at the edge of the Moon. Apart from these events, the main interest in observing occultations is the way in which the rapid motion of the Moon across the sky is dramatically revealed.

Planets

Like the Moon, the brighter planets are generally easy to find, especially if the observer has some familiarity with the constellations. Jupiter and Venus, near opposition and greatest elongation respectively, are far brighter even than Sirius, the brightest of the stars. Mars and Saturn are also bright enough to be readily found most of the time. Mercury, however, is quite a difficult object to locate. Although occasionally as bright as Sirius, it is always close to the Sun in the sky; its orbit is quite elliptical, so that its greatest elongation can vary considerably, but has a maximum values of only 28°. For this reason, Mercury is often also known as the evening or morning star, because it can be best seen soon after sunset when it is to the east of the Sun, and just before dawn when it is to the west of the Sun. In either case, it has to be observed against a bright twilight sky, and with the added haze and light pollution that is present at most urban or semi-urban observing sites, it then becomes almost impossible to see with the naked eye. Thus Mercury generally requires a telescope with accurate setting circles in order to be found. **If off-setting from the Sun, then all the precautions mentioned below in respect of solar observing must be followed.** Uranus is just bright enough to be seen with the naked eye at times. However, this would only be from a good observing site, under good conditions and for an observer with good eyesight. Neptune, Pluto and the asteroids are all too faint to be seen without a telescope, and so along with Uranus must usually be found from their tabulated positions using setting circles or by star hopping. The positions of the planets and the brighter asteroids may be found from several sources, and these are listed in Appendix 2.

The observer must be prepared for some disappointment on first observing a planet through a small telescope. This is for two reasons. Firstly articles and books on astronomy are often filled with magnificent pictures of the planets obtained from spacecraft. Not only are these obtained without the blurring effect of the Earth's atmosphere and from close to the planet, but also they are often computer-enhanced to make the colours and contrasts stronger. Sometimes, indeed, the image may be completely false-colour, either a monochromatic image with the different intensity levels being given different colours, or an image obtained partially or wholly in the ultraviolet or infrared and then visible colours substituted for those wavelengths. If such images are what the observer expects to see, then reality will be rather different. The second reason is that visually observing the planets requires a skill that can only be developed with practice. The skill required is that of taking advantage of the brief instants of clarity in the atmosphere to note the features of the planet. The turbulent effects of the atmosphere that degrade stellar images (see seeing, above) also affect planetary images. Since the planet is an extended object, the observed effect of seeing is slightly different from that on stars. When seen with the naked eye, planets are supposed not to twinkle in the manner of stars (though the author has never found this to be a reliable way of identifying planets). Through a telescope, each point making up the planetary image twinkles individually, and so the image as a whole is blurred and reduced in contrast. However, there will be occasional instants, lasting a fraction of a second to a few seconds, when the atmosphere is uniform along the line of sight. The planet will then be seen as clearly as the optics of the telescope allow. Thus, with practice, the observer can disregard the fuzzy image normally visible, and concentrate on what can be seen only under the best conditions. It is for this reason also that photographs or CCD images (Chapter 9) only rarely show the detail that can be determined by a skilled visual observer of the planets. Nonetheless, visual observations of the planets can be very rewarding. Even the inexperienced observer will find the sight of Saturn floating inside its incredible rings quite breathtaking. With experience, the observer will soon be able to distinguish the broad surface features and polar caps of Mars, details of Jupiter's spots and belts, transits and occultations of its Galilean satellites, the major divisions of Saturn's rings, and perhaps even check the disputed presence of light on the dark side of Venus (known as the ashen light).

Any type of telescope can be used to observe the planets, although some designs are likely to give better results than others. Features on the planets are of low contrast, and therefore the best telescope to use is one that degrades the available contrast least. Reflectors have secondary mirrors, and these are often supported by arms from the side of the telescope tube. Both the secondary mirror and its support arms affect the diffraction-limited image, broadening it and producing the spikes sometimes to be seen on images of stars (Figure 8.4). This has the effect of reducing the contrast of an extended object because each point of that object has an image in the shape of

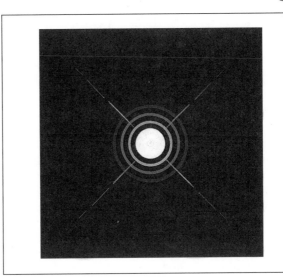

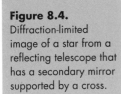

Figure 8.4.
Diffraction-limited
image of a star from a
reflecting telescope that
has a secondary mirror
supported by a cross.

that shown in Figure 8.4, and these then overlap to blur the image. Thus other things being equal, a Schmidt–Cassegrain or Maksutov telescope, which does not have support arms for its secondary, will be better for observing planets than a Cassegrain or Newtonian telescope, and a refractor, without any secondary mirror at all, will be better still. Other factors, of course, will influence the clarity of the image as well as that of the design of the telescope. Any factor that reduces contrast or increases the background illumination is deleterious. Thus the lowest magnification consistent with showing the required detail should be used and observations should be made when the Moon is not around. Open designs for telescope tubes (Figure 2.17) will usually be worse than closed tubes and should be covered if possible. Baffles should be carefully designed and positioned, all transmission surfaces should be anti-reflection coated, and all surfaces should be clean and dust-free. Serious planetary observers may even go to the extent of refiguring the optical surfaces of their telescope's objective to a far higher standard than usual. Improvements in the image are claimed to be found if surface accuracies of a twentieth of the oper-ating wavelength or better are used, rather than the more usual eighth of a wavelength.

Sun

CAUTION

The Sun should never be observed, either with the naked eye, or through a telescope, without the use of appropriate filters and stops or other devices designed to reduce its intensity in the visual and infrared parts of the spectrum. Only commercially produced filters designed for the type and size of telescope should be used. Smoked glass, exposed black and white film, CDs, space blankets, aluminised helium balloons, aluminised potato crisp packets, floppy disks, crossed Polaroid filters, smoked plastic, sunglasses, mirrors, X-ray photographs and so on are *not* suitable. Permanent damage to the eye and/or to the telescope can be caused if these precautions are ignored.

The observer should take the above warning to heart, but with the appropriate precautions should not let it stop him or her from observing the Sun. It is a most rewarding object, and has the not inconsiderable advantages of being observable in daytime and often under warm and pleasant conditions!

Finding

As the brightest object available to an astronomer, finding the Sun might not seem to be a problem. However, the warning given above applies just as much to looking through the finder telescope as through the main telescope, and even to sighting along the telescope tube with the naked eye to point it near the Sun. The finder can be used if it has an appropriate filter (see below). NB: Do not forget to cover the objective of the finder telescope if it does not have a filter. Alternatively, the setting circles can be used if the position of the Sun is known, though since the Sun moves quickly, the *Astronomical Almanac* (Appendix 2) will be needed to give its position. In the absence of either of these facilities, the Sun is best found by circularising the telescope's shadow; that is, the telescope is pointed very roughly towards the Sun, and then its shadow observed. As the telescope is moved, the size of the shadow of the tube will change, and be at its smallest when the telescope is pointing towards the Sun. For telescopes with a cylindrical tube, the shadow will then

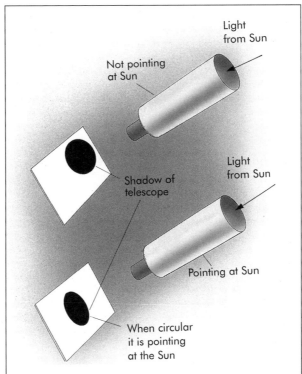

Figure 8.5. Finding the Sun by circularising the telescope's shadow.

be circular rather than elliptical (Figure 8.5). If done carefully, this procedure is accurate enough to bring the Sun into the field of view of the main telescope.

Observing

Full aperture filters are the best way of observing the Sun and are widely available from the telescope suppliers who advertise in popular astronomy journals (Appendices 1 and 2). These filters cover the objective and reduce the solar intensity before it becomes concentrated by the telescope optics. There is therefore no danger of damage to the telescope if, when finding the Sun, its image falls on to the mechanical parts of the instrument. This latter point is of considerable importance for Cassegrain, Schmidt–Cassegrain and Maksutov designs, because the focal ratio of the primary mirror in these is often quite small, leading to a primary image with great burning potential.

It is important to ensure that any filter used not only cuts down the visible light, but also eliminates the infrared part of the spectrum. If the image is too bright in the visible there is not much danger to the observer, because it will not be possible to look through the telescope for any length of time. However, a filter that cuts out the visible light, but not the infrared, will lead to the retina of the eye being burnt and to possible blindness, because there is no automatic physiological mechanism to protect the eye from excess infrared radiation. For safety, the solar surface brightness must be reduced by a factor of at least 30,000 (i.e. to 0.003%) across the whole spectrum.

A full aperture filter consists of a very thin film of plastic, covered by a thin deposit of aluminium. The whole film is thin enough to cause little distortion of the image, so that it does not need to be optically flat, and the aluminium reflects most of both the visible and infrared radiation. The reflectivity of aluminium falls off towards the blue, so images seen through such a filter have a slight bluish cast. Other metals are also used, resulting in other colours for the filtered Sun. Full aperture filters using optically flat glass are also produced, though they are generally more expensive than the metal-on-plastic type.

An alternative way of adapting a telescope to enable the Sun to be observed is via eyepiece projection. Note that eyepiece projection should *not* be used with telescope designs such as the Schmidt–Cassegrain or Maksutov where a short focal-ratio primary mirror forms a real image of the Sun inside the telescope. Indeed many manufacturers' guarantees will be voided if you use such telescopes for solar eyepiece projection. Always check with the manufacturer of your telescope before undertaking any solar work.

The telescope should be stopped down to about 50 mm if it is larger than this, by covering the objective using a securely attached piece of cardboard with a 50 mm hole cut into it. The telescope is then pointed at the Sun (see above), and a sheet of white cardboard placed 30–40 cm behind the eyepiece. The bright disc of the Sun will then be visible, projected on to this card (Figure 8.6), and the eyepiece position should be adjusted to bring the image into sharp focus. It will normally be helpful to attach additional cardboard around the telescope tube to act as shielding.

Filters at one time were produced that could be placed directly into the eyepiece. However, these

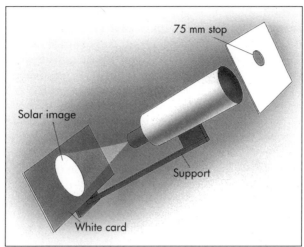

Figure 8.6.
Observing the Sun by
eyepiece projection.

should be avoided for several reasons: firstly, the filter
will get very hot and the thermal stress may cause it to
shatter, allowing the undiluted radiation from the Sun
to enter the eye; secondly the heat may damage the eyepiece; and thirdly the heat from the filter will lead to
convection currents and possibly distort the optics of
the eyepiece, ruining the image quality.

Specialised solar telescopes at major observatories
are often fixed and the light is then fed into them by a
coelostat. This is a device that has an arrangement of
two plane mirrors, both on driven mountings, so that
the light from any part of the sky can be reflected
towards a fixed direction, and the object (usually the
Sun) tracked as it moves across the sky.

With any of the above methods, it should be possible
to observe limb darkening, sunspots, plages, etc. and to
measure the solar rotation period, which varies with
solar latitude. Sunspot counts can easily be made in
order to monitor the activity of the Sun and to observe
the 11-year solar cycle, though some practice is
required to obtain consistency. The Zurich sunspot
number, R, is obtained from the formula

$$R = k(10g + s) \tag{8.4}$$

where g is the number of visible sunspot regions, either
groups or single spots, s is the number of individual
sunspots (both the isolated spots and those within the
groups) and k is a personal correction factor to take
account of the instrument used, the quality of the observing site, and the idiosyncrasies of the observer in

allocating spots into groups. It has to be determined by comparing the sunspot counts obtained by the observer over a period of a few months with the published values.

At a very considerably higher cost than the full aperture filters just considered, H-α filters are obtainable. These are very narrow-band filters centred on the red line of hydrogen at 656 nm; they are primarily intended to enable prominences to be observed at the edge of the Sun, and filaments, flares, etc. on its disc.

A reasonably skilled DIY enthusiast can make an alternative device for observing prominences for little cost. This is the prominence spectroscope. It is a spectroscope with an additional slit centred on one of the strong solar absorption lines, usually H-α. The entrance slit to the spectroscope is aligned on the limb of the Sun, and the observer looks through the second slit. Prominences may then be seen because the prominence has an emission line spectrum and so emits strongly at that wavelength, while the solar photospheric emission is spread over the whole spectrum.

Stars

Stars have been observed by astronomers since before recorded history. Only in the last few centuries, however, have they been observed in their own right; before that they were just a background against which the Sun, Moon and planets moved. The two basic visual observations possible for stars are of their positions and brightnesses. Positions have been considered in Chapters 4 and 5; here therefore we should look at stellar brightnesses. Firstly, however, let us try and reduce the confusion that many beginners experience over the apparently haphazard naming of the stars.

Stellar Nomenclature

The naming of stars appears to be haphazard because of the many different systems that are in use at the same time. Some stars, especially the fainter ones, may only have one label, but most of the brighter stars can have several names. Thus Sirius is also known as α CMa, 9 CMa, GSC 0594902767, HIP 32349, HR 2491, BS 2491, ADS 5423, HD 48915, and SAO 151881 amongst many other possibilities.

Half a millennium ago, stars were identified by being given proper names like Betelgeuse, Albireo or Deneb, or their position in the sky described, as in "The left-hand star of the 'W' of Cassiopeia". The latter method was cumbersome, and neither method was systematic. The first systematic method of naming stars originated in the star catalogue *Uranometria* published in 1603 by Johan Bayer (1572–1625). The Bayer system used a Greek letter in the order of the stars' brightnesses within the constellation followed by the constellation name. For example the previously named stars become α CMA (Sirius), α Ori (Betelgeuse), β Cyg (Albireo), and α Cyg (Deneb). There are often anomalies to the system arising from changes to the constellations, or mistakes, with some constellations missing some letters, or brighter stars having later letters than fainter ones, etc. Thus α Orionis (0.9^m) is fainter than β Ori (0.1^m). After the Greek letters have been used, then the lower case letters, a, b, c, ... may be used in a similar way, and after the lower case letters, upper case letters from A to P. The letters from R to Z are reserved for variable stars (see below).

A second systematic method of identifying stars, the Flamsteed number, originated in John Flamsteed's (1646–1719) catalogue published in 1725. The stars are numbered within each constellation in order of increasing right ascension. For example, 57 UMa, 88 Leo or 15 Pic.

Over the last hundred years, thousands of star catalogues have been produced, many containing just a few stars of a particular type but some with hundreds of thousands or millions or more stars. The second version of the Hubble Space Telescope Guide Star Catalogue (GSC 2) is expected out in early 2003 and will list half a billion objects. Many stars are therefore identified by their numbers in one or more of these catalogues. The label then usually takes the form of an abbreviation of the catalogue's name followed by the star's number. Commonly encountered usages of this are based upon the Henry Draper Catalogue (HD), the Bright Star Catalogue (BS), the Smithsonian Astrophysical Observatory Catalogue (SAO), the Hipparcos Catalogue (HIP) and the Hubble Space Telescope Guide Star Catalogue version 1.2 (GSC). A slight exception to this practice occurs with the Bonner Durchmusterung and Córdoba Durchmusterung Catalogues (BD and CD). These catalogues are subdivided by the stars' declinations. The label therefore takes the form of the declination followed by a running number, e.g. BD +12° 1234.

Variable stars may well be named within one or more of the above systems. The well-known eclipsing binary, Algol, for example, is also β Per, 26 Per, BS 936, HD19356, SAO38592, etc. However, there are also several systems specifically designed for naming variable stars. An extension to the Bayer system is the most widespread approach. This labels the variables in each constellation with capital letters, starting at R for the brightest. Letters from R to Z suffice for only nine variables in each constellation, and so thereafter double capitals are used. RR to RZ is used for the next nine variables discovered, then SS to SZ for the next eight, TT to TZ for the next seven, and so on to ZZ, giving another 54 labels in all. After that the sequences AA to AZ, BB to BZ, CC to CZ, …, QQ to QZ are used. A total of 334 variables in each constellation may thus be identified, with the letters followed by the constellation abbreviation; S Cyg, RR Lyr, FG Sag, etc.

When more than 334 variables are found the simpler practice of using the letter V followed by a number starting from 335 is used. Thus we have V861 Sco, V444 Cyg, and V432 Her, etc.

Pulsars and other special groups of variables may use other systems of labelling; a common practice is to use a label to indicate the variable type and combine it with an approximate right ascension and declination. Thus we have the pulsar, PSR 1257+12, which is to be found near RA 12h 57m and Dec +12°. Gamma ray bursters are named for the date of their occurrence as GRB yymmdd. Thus GRB 970228 occurred on 28 February 1997.

It is thus not surprising that the naming of stars appears haphazard at first (or even second) sight. However, it *does* all become familiar after a while! It is worth adding, though, that with only a few exceptions, such as Barnard's star, stars are *not* named after people.[5] The numerous adverts that one sees in the

---

[5] Comets and asteroids can be named after people, but you have to discover a new one yourself to have the right to name it. On discovery, comets are designated by the year of discovery and a running letter plus, usually, the discoverer's name (e.g. comet Kohoutek 1973f). Once an orbit has been determined, then the name changes to the year of perihelion passage plus a running Roman numeral giving its order of perihelion passage (e.g. comet West 1976 VI). The discovery of a new asteroid, after which you are entitled to name it subject to certain restrictions, such as not using an existing name, is somewhat more complex. Firstly you have to obtain at least three separate positions for the object so that its orbital parameters can be determined. It then has to be retrieved at two successive oppositions before it can be added to the list of known asteroids.

popular press purporting to name stars after you for the payment of a fee are therefore spurious. The operators of such scheme will doubtless write your name against your chosen star on their star map, but it will not be officially recognised and no one else will use it. So do not waste your money.

Magnitudes

The first known eye estimates of the brightness of stars occurred in Hipparchus' star catalogue of about 150 BC, though there had been at least one earlier catalogue produced by Timocharis and Aristyllus around 300 BC. Hipparchus divided the thousand or so stars in his catalogue into six groups based upon their brightnesses, with group I being the brightest, and group 6 those only just visible. This classification remained adequate until the invention of the telescope revealed stars fainter than those visible to the naked eye. Such stars were then placed into groups 7, 8, 9 and so on. By the middle of the nineteenth century, however, more precise measurements of stellar brightnesses (photometry, see Chapter 11) were being pioneered by William Herschel's son John (1792–1871) and others. Now Hipparchus' grouping of stars according to their brightnesses tried to divide the total range in a uniform manner. But the response of the eye to intensity is geometric rather than arithmetic – that is, if the eye perceives stars A and B to differ by a similar amount to the difference between stars C and D, then it will be the *ratio* of the energy from A to that from B that will be similar to the *ratio* of the energy from C to that from D, and not their energy *differences*. This is known as the Weber–Fechner law. Thus the difference from one of Hipparchus' groups to the next was a constant factor in energy terms, not a constant difference. This was codified in 1856 by Norman Pogson (1829–91) into the system of stellar magnitudes still in use today. The system is defined by Pogson's equation:

$$m_1 - m_2 = -2.5\log_{10}\left[\frac{E_1}{E_2}\right] \qquad (8.5)$$

where m_1 is the magnitude of the first star, and E_1 its energy, m_2 is the magnitude of the second star, and E_2 its energy.

Equation (8.5) gives a constant factor in energy terms of $\times 2.512 \ldots (= 10^{0.4})$ between one magnitude and the next as required, and was chosen to ensure that measurements under the new system corresponded as closely as possible to those under the previous groupings. The equation is a relative one, however, and gives the magnitude of one star with respect to that of another. The magnitude scale must therefore be calibrated by assigning an arbitrary value for the magnitude of at least one star (known as a standard star). The magnitudes of all other stars will then follow from Equation (8.5) by comparison with that standard. In practice, there are many standard stars distributed over the whole sky, so that there is always one reasonably close to any star that is being observed. The magnitudes assigned to the standards are again chosen so that the magnitudes correspond as closely as possible to the previous estimates, and are such that a star of magnitude 6 should just be visible to the unaided eye. This calibration results in a star of magnitude 1 having an energy (irradiance) of 9.87×10^{-9} W m^{-2}, at the top of the Earth's atmosphere.

Because of these historical roots to the stellar magnitude scale, it is rather awkward to use. Thus against all normal practice, the *brighter* the star, the *smaller* the resulting number (magnitude). Setting magnitude 6 stars to be those just visible results in the brightest objects being outside the range, and so it has to be extended to zero and negative magnitudes; Sirius, for example, has a magnitude of –1.45. Since it is a logarithmic scale it is very compressed, the range from the brightest object in the sky (the Sun) to the faintest detectable in the largest telescopes being only about 55 magnitudes, but this corresponds to an actual difference in their energies of a factor of 10^{22}!

The relationship between magnitude difference and energy difference is tabulated in Table 8.1, and some example magnitudes listed in Table 8.2. The magnitudes listed in that table are all apparent magnitudes, that is, the magnitude of the star or other object as it appears in the sky, and therefore the important quantity to know when observing the object. Apparent magnitudes, however, take no account of the distance of the object, and so are unrelated to the object's actual luminosity. Thus Sirius is apparently twice as bright as Betelgeuse, but because Betelgeuse is much further away, it is actually over 900 times brighter than Sirius. A second magnitude scale is therefore also used which takes account of distance. This is the absolute magnitude, and is discussed in Chapter 11.

Table 8.1. Magnitude/energy relationship

Δm	E_1/E_2
0.1	1.1
0.2	1.2
0.3	1.3
0.4	1.4
0.5	1.6
0.6	1.7
0.7	1.9
0.8	2.1
0.9	2.3
1.0	2.5
2.0	6.3
3.0	15.9
4.0	39.8
5.0	100 (exact)
10.0	10 000 (exact)
15.0	1000 000 (exact)
20.0	1000 000 000 (exact)
50.0	100 000 000 000 000 000 000 (exact)

Observing Stars

For stellar observations there is no substitute for aperture – the bigger the better! This is because for point sources (Equation 2.2), unlike extended sources, the telescope does increase the brightness. From the

Table 8.2. Examples of apparent magnitudes

Object	m
Sun	−26.7
Full Moon	−12.7
Venus (maximum)	−4.3
Jupiter (maximum)	−2.6
Mars (maximum)	−2.02
Sirius A	−1.45
Betelgeuse	−0.73
Polaris	+2.3
Uranus (maximum)	+5.5
Faintest object visible to the unaided eye	+6.0
Faintest object visible in a 0.3 m telescope	+13
Faintest object visible in a 1 m telescope	+16
Faintest object visible in a 10 m telescope	+21
Faintest object detectable using the very best of modern techniques	+28

definition of stellar magnitudes, a star of magnitude 6 is one that should just be visible to the unaided eye. However, that would be from a good observing site, under good conditions and for an observer with good eyesight. Magnitude 5 is therefore often a more realistic limiting magnitude for unaided observations, and from a poor site it could be considerably worse. We thus get as a realistic limiting magnitude for a particular telescope used visually, roughly that

$$m_{\mathrm{lim}} = 16 + 5\log_{10} D \qquad (8.6)$$

where D is the objective diameter in metres.

Visual observations of a single star can give information on its brightness, and in a few cases, its colour. The brightness can be estimated if another star of known magnitude is within the field of view, and with practice an accuracy of ± 0.1 or ± 0.2 magnitude can be achieved. This is very useful for observing and monitoring variable stars, especially irregular variables and the explosive variables like novae. An eye estimate will reveal whether or not the star has changed appreciably from a previous observation, and may be of sufficient accuracy in itself, or indicate whether or not it is worth making a more precise measurement (Chapter 11). A few stars are bright enough to trigger the colour vision of the eye. Thus Antares (α Sco) is clearly red, and Sirius (α UMa) and Vega (α Lyr) bluish-white, even to the naked eye. Even more spectacular are double stars with components of widely differing temperatures so that the colours show up strongly by contrast. A few of the brighter examples of these are listed in Table 8.3.

Table 8.3. Double stars with strong colour contrasts

Star	Colours (brightest star first)
γ And	Red, green and blue
ε Boo	Yellow and blue
ξ Boo	Orange and red
α CVn	Yellow and blue
ι Cas	Yellow, red and green
β Cyg	Yellow and blue
γ Del	Yellow and green
α Her	Orange and green
β Ori	Blue-white and blue
η Per	Red and blue
α Sco	Red and green
β Sco	Green and blue

Much of the interest in visually observing stars, however, is not for the single stars, but when two or more are closely associated. This can range from just a pair of stars to very large groupings, called globular clusters, containing up to a million members. The latter are really small galaxies and are considered under that heading below. Pairs of stars are divided into double stars and binary stars. There is no difference between the two in their visual appearance, but in a double star the two stars happen to lie along the same line of sight, but are actually separated in space by a very large distance. In a binary, the two stars are physically close together as well as apparently close in the sky, and one will usually be in orbit around the other.

Most visual binaries have very long orbital periods, but in some cases the orbital motion can be detected over a period of a few years. The relative positions of the two stars will need to be measured accurately if the orbital motion is to be found, and a micrometer eyepiece will be needed for this purpose (Chapter 2). The separation of the binary is measured by aligning the horizontal cross-wire between the two stars, moving the telescope to set the intersection of the fixed cross-wires on to one star, and adjusting the moving cross-wire to set it on to the other star (Figure 8.7). The scale of the micrometer can be calibrated from the rate of drift when the telescope drive is turned off (Equations 8.1 and 8.2). The position angle is the angle from the north direction measured in the sense N-E-S-W, and can be read directly from the angular scale of the micrometer (Figure 8.7), with the east–west direction as seen in the eyepiece being established from the drift direction. A few of the brighter visual binaries are listed in Table 8.4.

Both double and binary stars are useful as tests of the performance of a telescope, the skill of the observer, and the quality of the observing conditions. The diffraction-limited resolution of a telescope given by Equations (2.3) and (2.4) is for two stars of equal brightness. It is also an arbitrary though generally realistic definition. But an experienced observer can sometimes detect that the image of a double star with a separation of only a half or a third of the diffraction limit appears to be different from the image of a single star, and it has therefore in some sense been resolved. Conversely, for stars of widely differing magnitudes, the separation may need to be ten or more times the theoretical resolution before the stars can actually be split.

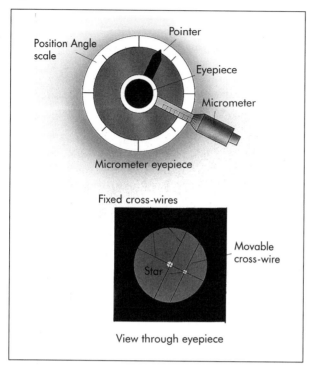

Figure 8.7.
Micrometer eyepiece.

Remembering that high magnifications are needed to reach the diffraction limit (Equation 2.19), observations of close double stars under the best observing conditions will thus test the quality of the telescope; under poorer observing conditions, the closest double star separable will provide a measure of how good or bad those conditions may be. Lists of double stars may be found in many sources (Appendix 2).

Multi-star asterisms ranging from triplets to galactic clusters like the Pleiades containing hundreds of stars are also to be found, though less widespread than doubles and binaries. Many are listed in the same

Table 8.4. Visual binary stars

Star	Magnitudes		Period (years)
ε Boo	4.7	6.7	150
α CMa (Sirius)	−1.5	8.6	50
η Cas	3.4	7.2	480
α Cen	0.1	1.4	80
ζ Her	2.9	5.5	34
70 Oph	5.1	8.5	88
85 Peg	5.8	8.9	26

sources as the double stars (Appendix 2). They may again just be chance alignments or be physically linked, but as the number of stars in the asterism increases, the probability of there being a physical association rises rapidly. Similar observations may be made for these objects as those described for the single and double stars. The larger and denser clusters, however, are also well worth finding and observing for their beauty alone.

Nebulae and Galaxies

Observations of diffuse interstellar nebulae (H II regions, supernova remnants, planetary nebulae, absorption nebulae, etc.), star clouds, globular clusters and galaxies have much in common with each other. Despite the very different physical natures of the objects, in the telescope their visual appearance is as faint diffuse extended objects. It is worth repeating here the warning given in respect of observing the planets – that images of these objects to be found in books, etc. far exceed their visual appearances in small telescopes. Even the brightest of them hardly trigger colour vision in the eye (remember a telescope does not increase the surface brightness of an extended object, Equation (2.16)), and so they will appear milky white, not resplendent in all their spectacular colours. Furthermore, the images in books are the result of long exposures, sometimes with CCD and other detectors which have a hundred or more times the sensitivity of the eye, so that they reveal far more of the objects' extents than can be seen visually. Nonetheless, observations of these objects can still be fascinating and highly rewarding. With galaxies, one is looking far into the depths of the universe. With a 0.3 m telescope, for example, the limiting magnitude for a distant galaxy can range from magnitude 9 to 12 depending upon the galaxy's angular size and the sensitivity of the observer's eyes (not magnitude 13 as given by Equation (8.6), because of the extended nature of the galaxy). Thus galaxies can be seen directly that are over 200 million light years away, and the photons actually entering the eye have then been travelling through space since before the time during which dinosaurs roamed the Earth. H II regions, like the Orion nebula, are the birthplaces of stars and planets, and supernova remnants, like the Crab nebula, their most spectacular death throes.

Given the faintness and low contrast of many of these objects, observing them is one of the most severe tests possible of an observer's skill. The lowest-power eyepiece consistent with showing the object as extended should always be used. The optics of the telescope and the eyepiece should be as clean and dust-free as possible. Any trace of dew will be fatal to finding even the brightest nebulae or galaxies. Except for the very brightest, such as the Orion nebula or the Andromeda galaxy, these objects have to be found from their tabulated positions and the use of accurate setting circles, or by star hopping from a nearby bright star. Even when the telescope is pointing at the object, it may be necessary to use averted vision (Chapter 2) to see it at all. For really difficult objects, it can be helpful to nudge the telescope back and forth by small amounts, and the object may then be found to show up while moving, even though it disappears again when stationary.

Lists of extended objects and their positions may be found in many places (Appendix 2). Two catalogues in particular, however, are to be noted because they are used to give names to many extended objects: these are the *Messier list* and the *New General Catalogue.* Charles Messier (1730–1817) was a comet hunter who got exasperated with continually mistaking nebulae and galaxies for comets, and therefore produced a list of such objects in order to avoid them! There were 103 objects in Messier's original list, but one was a double star and another either a duplication or even a comet. Later additions to the list have increased the number to 110, of which 39 are galaxies, 29 globular clusters, 27 galactic clusters, 11 gaseous nebulae, and one each a double star, an asterism and a bright patch of the Milky Way. The Messier objects are identified by an "M" and a number from the catalogue, Thus the Crab nebula is M1, the Orion nebula M42, the Pleiades M45, and the Andromeda galaxy M31. For the observer using a small telescope, the Messier list is useful since it contains many of the extended objects likely to be observable. It can also be used as a challenge; initially just to try and observe as many of the objects as possible. M74 is probably the most difficult of the Messier objects, so if you can find it, you should be able to see all the others. With more experience, the challenge can be to find as many Messier objects in a single observing session as possible, and this can be particularly interesting as a competition among members of an astronomy society or at a "star party", etc.

The second of the catalogues originated in 1786 as William Herschel's *Catalogue of Nebulae,* and was extended by his son John and then by Johann Dreyer in 1888 into the *New General Catalogue of Nebulas and Clusters of Stars* containing nearly 8000 objects. This catalogue is abbreviated as NGC, and objects within it are referenced by this identifier and their catalogue number. Thus under this system of nomenclature, the Crab nebula is NGC1952, the Orion nebula NGC1976, and the Andromeda galaxy NGC224. The Pleiades are not listed in the NGC catalogue since the cluster is close enough to us for the member stars to be observed individually, though other galactic clusters are included.

Recently a new catalogue, the Caldwell catalogue, produced by Sir Patrick Moore (whose full surname is Caldwell-Moore), has found some favour. The catalogue mirrors the Messier catalogue in that it also has 109 objects within it. The Caldwell catalogue, however, covers the whole sky and the running numbers increase consistently from north to north. C plus the running number designate objects; for example, the star cluster h and χ Per is C14.

Daytime Observing

Many people will have noticed that the Moon is often visible to the naked eye during the daytime, and can obviously therefore also be observed through a telescope. The brighter planets and stars can also often be found. For any daytime observing, the prime requirement is for a haze-free sky to reduce the background light. This requirement does not necessarily mean cloud-free. The most transparent skies often occur after a rainstorm when the dust has been washed out of the atmosphere. The gaps between the clouds can then be used on an opportunistic basis for observing. Allied to this, the further the object is away from the Sun in the sky, generally the easier it will be to observe. The scattered solar light (i.e. the blue sky) is polarised, approaching 100% linear polarisation at 90° from the Sun under good conditions. A sheet of Polaroid, aligned orthogonally to the sky polarisation, can therefore help to reduce the background light considerably. Objects have to be found either from their positions and the use of setting circles, or by off-setting from the Sun (**note that all the precautions for observing the Sun discussed earlier must be employed, even when it is only to be used as the starting point**). An important point to watch is that, if the telescope is even slightly out of focus, then the objects will become impossible to see. If the eyepiece focusing mount has a position scale, a note should be made of its reading when in focus. If, as is more usual, the focusing mount is not calibrated, the telescope should be focused at night and left undisturbed so that it is still in focus the next day. If this is not possible, then the telescope can be focused on a terrestrial object on the horizon provided that it is at least several hundred metres away, and the focus corrected to the position for an infinite distance using the formula

$$\text{Focal correction} = \frac{1000 f^2}{d - f} \text{ mm} \qquad (8.7)$$

where the focal correction is the distance that the eyepiece must be moved towards the objective, f is the focal length of the telescope in metres and d is the approximate distance of the terrestrial object in metres.

This last approach, however, is not very satisfactory because even for an object a kilometre away, the focal correction is over 4 mm for a 0.2 m f10 telescope, and therefore difficult to determine precisely in the absence of a position scale.

False Observations

For a variety of reasons, observations can be misleading, or even mistaken. This comment applies to all types of observing, but is particularly apposite for visual observations since there is then no hard copy that can be checked later or independently. Problems are most likely to occur when the observer is trying to observe at the limits of his or her equipment and his or her abilities. Sometimes the mistake can result from preconceived ideas, leading to the observer seeing what he or she expects to see, and not what is actually there. At other times, the false observation can arise from the physiological nature of the human eye. Thus several

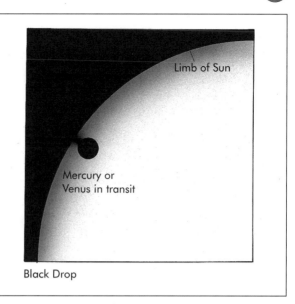

Figure 8.8.
"Teardrop" effect during a transit of Mercury or Venus. The silhouette of the planet still appears connected to the limb of the Sun, even when there should be a thin thread of the photosphere visible.

objects on or just below the resolution limit of the eye may be seen linked together to form a linear feature. This is probably what happened with the Martian canals. They were seen, not because the observers were poor, but because their eyesight was so acute that they were almost able to see the volcanoes and craters that we now know actually to be on the Martian surface. Many of the rods and cones in the eye's retina have cross-links, and this can lead to problems when high levels of contrast occur in an image. The "tear-drop" effect (Figure 8.8) found when Mercury or Venus transit the Sun is an example of this. There is little an observer can do to counteract the possibility of false observations, except to be aware that they can occur, and to be cautious about any results obtained towards the limits of what may be possible.

Exercise

8.1 Calculate the energy per square metre at the surface of the Earth from a star just visible to the naked eye from a good observing site (ignore atmospheric absorption). (Irradiance from a 0 magnitude star at the top of the Earth's atmosphere: 2.48×10^{-8} W m^{-2}.)

Chapter 9

Detectors and Imaging

The Eye

The front line detectors for almost all astronomers are their own eyes. For many, especially when using smaller telescopes, these are also the only detectors. The eye, or more particularly vision, which is the result of the eye and brain acting in concert, is, however, a very complex phenomenon, and some knowledge of its peculiarities is essential for the observer. Thus reference has already been made in Chapter 8 to averted vision, the effect of high contrasts (known as irradiation), and the combination of subresolution features (Martian canals). The structure of the whole eye (Figure 9.1) is well known from school, and need not be considered further here. It is the structure of the eye's detector, the retina, that is of importance.

The retina is a network of detecting cells connected by nerve fibres via the optic nerve to the brain. The detecting cells are of four types: rods, and three types of cones. The names of the cells come from their shapes. The three types of cones have sensitivities which peak at about 430 nm, 520 nm and 580 nm, and they provide us with colour vision. The overall sensitivity of the cones is, however, low. The rods are of only one type, and have a response that peaks at 510 nm. There are about 10^8 rods and 6×10^6 cones but only 10^6 nerve fibres; hence many cells are linked to each nerve. Cones are most abundant in the *fovea centralis* and many there are singly connected to nerve fibres. The *fovea centralis* is the point on the retina where the light

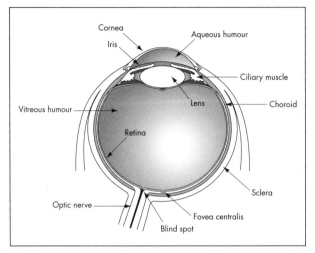

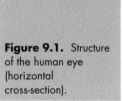

Figure 9.1. Structure of the human eye (horizontal cross-section).

falls if we look directly at an object; the rods become commoner and the cones fewer as distance increases away from this point.

The light-sensitive component of the rods is a molecule called rhodopsin or visual purple (from its colour). Under high light levels, the rhodopsin is mostly inactivated and the rods have a low sensitivity. The cones then dominate vision and we see things in colour. At low levels of illumination, the rhodopsin regenerates over a period of 20–30 minutes, restoring the rods to their full sensitivity. Vision is then almost entirely via the rods. This effect has two consequences for astronomers. The first is the familiar effect of dark adaptation: immediately on leaving a brightly lit area into the dark, very little can be distinguished. However, after a few minutes, vision becomes much clearer, and after half an hour or so one can often see quite easily even on a moonless night. This adaptation is partly a result of the pupil of the eye increasing in size, but much more the effect of the regeneration of the rhodopsin, which increases the retina's sensitivity by a factor of 100 or more. The rhodopsin is destroyed quickly by strong light, and so the dark adaptation can easily be lost if the observer is careless with a torch or turns on the main lights, etc. However, the sensitivity of rhodopsin ranges from about 380 nm to 600 nm. So deep red light, with a wavelength longer than 600 nm, should not affect dark adaptation because the rhodopsin will not absorb that radiation. For this reason, many observatories are illuminated with red lights to allow ease of movement while preserving dark adaptation. It should be stressed, however, that for this to work the light has to be very red indeed, with no short-wave leakages. The second effect of rod vision is that of lack of colour. There is only the one type of rod, and so at low light levels the eye cannot distinguish between different wavelengths. Thus, as already mentioned (Chapter 8), most gaseous nebulae appear milky white to the eye, instead of the multi-hued reds, greens and blues to be seen on many photographs.

The integrated sensitivity of the cones has a peak near 550 nm, while the sensitivity of the rods peaks near 510 nm. If a bright star and a faint star are simultaneously in the field of view of a telescope, the former may be seen with colour vision while the latter is detected by the rods. The two stars will thus be seen in slightly different parts of the spectrum, and this can lead to incorrect estimates of their relative magnitudes, with a faint blue star being seen as too bright, and a bright red star as too faint, a phenomenon known as the Purkinje effect.

Reference has been made elsewhere to other properties of the eye that affect astronomical observation:

1. The eye's logarithmic response to intensity, which results in the stellar magnitude scale (Equation 8.5).

2. Averted vision, which works because rods increase in abundance away from the *fovea centralis,* and looking to the side of the object therefore causes its light to fall on to the higher-sensitivity rods.

3. High contrast effects (the tear-drop effect, etc. – Figure 8.8), which arise because of the interconnection of many cells to a single nerve. If only a few of those cells are stimulated, then the nerve does not trigger, and so the brain registers that part of the image as dark, etc.

4. The resolution of the eye, which is 3′ to 5′, instead of the 20″ suggested by Equation (2.4), and this is due to aberrations arising from the poor optical quality of the eye, the granular nature of the retina because it is composed of individual rods and cones, and the interconnection of many rods and cones to a single nerve.

Hence these need not be considered further here.

Charge Coupled Devices (CCDs)

Many observers are content just to look through their telescopes, especially when starting in astronomy. Sooner or later, however, most people want to obtain a permanent record of what they can see through the telescope. This may, of course, be in order to boast to friends and rivals in an astronomy society, or perhaps to enable more rigorous measurements and serious work to be undertaken. There are many types of detectors sensitive to light, but only three are likely to be of interest to observers with small to medium-sized telescopes; these are the charge coupled device (CCD), the photographic emulsion, and the p-i-n photodiode, and they are each considered in turn.

The CCD has been in use on major telescopes since the early 1970s, and for most purposes has completely replaced the photographic emulsion and other types of detector. Only in the last few years, however, has it become cheap enough to be considered outside the professional observatories. Even now, "cheap" is a relative term, since a CCD will need a small computer, image processing software and other accessories, and the observer will (at current prices in 2003) need to pay about half the cost of an entire 0.2 m Schmidt–Cassegrain telescope for a working CCD package with around 1000×1000 pixels. The CCD, however, is about 100 times as sensitive as either the eye or the photographic emulsion, and so a 0.2 m telescope used with a CCD becomes the equivalent of a 2 m telescope used visually or photographically! The cost of the CCD considered in this fashion thus becomes a huge bargain.

Anyone purchasing a CCD will find detailed instructions on its use given by the manufacturer. Here, therefore, we only look at the basic operating principles.

Consider first an array of atoms, perhaps forming part of a crystal, and each with one electron loosely attached (Figure 9.2). The absence of an electron in such an array (Figure 9.3) leaves a hole that, because of the absence of the negative charge of the electron, will then appear to be a positive charge. Applying a positive voltage from, say, the right, will cause the electron to the left of the hole to hop into that hole, leaving a new hole in its place one unit to the left (Figure 9.4). The next left-most electron then hops into that hole, and a

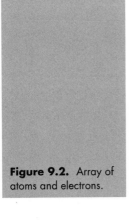

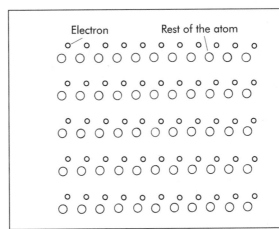

Figure 9.2. Array of atoms and electrons.

new hole appears one unit further to the left, and so on. Thus effectively the hole behaves as though it is a positively charged particle, and for most purposes we can treat it exactly as though it is such a particle.

The detection mechanism of a CCD is the production of electron–hole pairs in a crystal of silicon. In the CCD, the absorption of a photon in p-type silicon "frees" an electron from the grid, so that it can wander around within the solid (the explanation for this in terms of energy bands within a solid is beyond the scope of this book, but the interested reader is referred to any book on condensed matter physics). A mobile positive hole, as above, is also produced by the absence of this electron from its "normal" place. The electron and hole will move at random within the crystal under the effects of the thermal motions of the atoms, and if

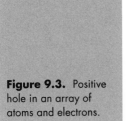

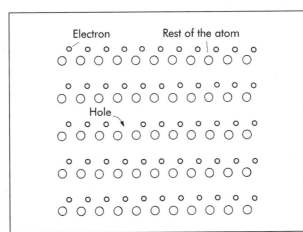

Figure 9.3. Positive hole in an array of atoms and electrons.

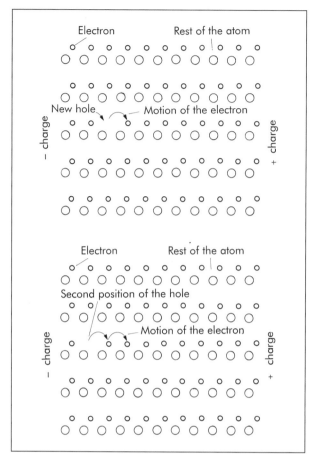

Figure 9.4. Motion of electrons and a hole in an array of atoms and electrons.

they should encounter each other, they will recombine and emit a photon. There will then have been no net detection of the original photon. Thus the electron and the hole must be kept apart. In the CCD, the electron is attracted towards a positively charged electrode placed on, but insulated from, the silicon. Simultaneously, the hole is repelled by the same charge (Figure 9.5). The electron is then stored beneath the electrode, where it will subsequently be joined by other electrons produced in the same manner by other photons. The hole is ejected into the body of the silicon, and eventually lost. Thus a negative charge builds up under the electrode, whose magnitude is proportional to the intensity of the illuminating radiation.

Many such basic units can be produced in a two-dimensional array by standard very large integrated circuit (VLSI) techniques (Figure 9.6), to enable a two-

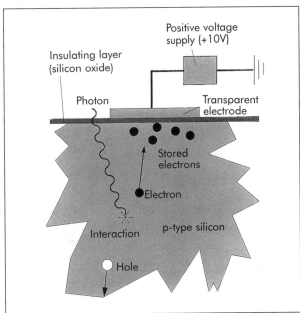

Figure 9.5. Basic unit of a CCD.

Figure 9.6. CCD array (schematic); in practice, CCDs can have up to 10 000 × 10 000 elements.

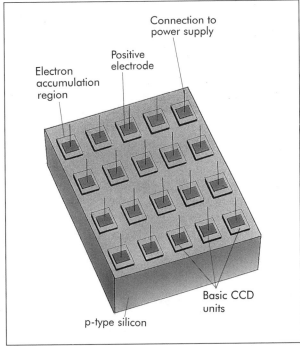

dimensional charge analogue of the incoming photon image to be built up. The image is read out from the silicon chip by moving (coupling) the charge under the first electrode to an output electrode, from whence it may be detected. The charge under the second electrode is then moved to the first, and thence to the output. The charge under the third electrode is moved to the second and is then moved to the first, and thence to the output, and so on. The image is thus output as a series of charge packets, each of whose magnitudes is proportional to the intensity of the image at the point in the two-dimensional array from which the charge originated. This may then be displayed as a video image if read at 50 Hz or 60 Hz, or output to a computer, etc. if longer exposures are used.

For astronomical purposes, the CCD has to be cooled to reduce noise and inter-pixel isolation breakdown, and exposures can then be for many hours. The efficiency of detection of the CCD can reach 80% at wavelengths of about 700–800 nm, making it up to 100 times faster than photographic emulsion and the human eye. The basic CCD has a spectral response extending from about 450 nm to 1000 nm. This can be extended to shorter wavelengths by a suitable fluorescent coating to convert the shorter-wavelength light into that detectable by the CCD. After its efficiency, the other main advantage of the CCD is that its output can be fed directly into a computer. The CCD's disadvantages are its high cost and small size compared with photographic emulsion.

For professional use, devices with up to about 10 000 × 10 000 pixels are available. These latter are very expensive, and since they are usually thinned to about 10 µm, so that they can be illuminated from the back in order to improve their sensitivity, they are also fragile. In addition, processing such large images requires substantial computing power and specialist software, well beyond the capacity of currently available PCs.

Digital cameras with 3 or 4 million pixels are now available for around £500 (US$800), and these of course use CCDs to obtain their images. While such cameras can be attached to telescopes and produce excellent images, they cannot offer long exposures since the CCDs within them are not cooled.

Image Processing

One of the advantages of a CCD image is that it may be fed straight into a computer for image processing, and normally this advantage is also a necessity. The CCD image has to be corrected in various ways, if optimum extraction of information from it is to be achieved (all images, howsoever obtained, may benefit from or need image processing – it is mentioned here in connection with CCD images only because they are already stored on a computer and can therefore easily be processed).

- Flat fielding – the relative responses of all the elements may vary. An exposure of a uniform field of view is therefore required and must be subtracted from the image to provide a "flat field".

- Cosmic rays – these are troublesome on long exposures. Each cosmic ray provides a bright "spike" in the image, which may be mistaken for a feature. Long exposure images must therefore be inspected carefully and such features removed; they are usually so much brighter than the rest of the image that they are easily recognisable. If need be, the affected pixels can be replaced by the average of their surroundings to provide a corrected image.

- Contrast stretching – a CCD can resolve 100 000 or more grey levels, while a typical computer display screen can only show 256 grey levels. A big improvement in usefulness may therefore be obtained by mapping the main grey levels in the image to those available on the display. Suppose, for example, that 90% of the data in an image is between grey levels 1200 and 3760. By setting levels below 1200 to display as level 0 for the monitor, levels 1200 and 1209 to display as level 1 for the monitor, levels 1210 and 1219 to display as level 2 for the monitor, levels 1220 and 1229 to display as level 3 for the monitor, and so on, then optimum use may be made of the available information. Such a remapping of the image is called grey scaling or contrast stretching.

- Background and dark signal subtraction – thermal and other noise sources will result in electrons accumulating in the CCD even when it is completely in the dark. The sky background, especially from urban sites, can be quite bright. Both of these problems can be reduced by taking additional exposures with the CCD, in the first case with the shutter closed, and in the second case while looking at a featureless part of the sky near the object of interest. These two exposures are then subtracted from the main exposure to reduce the noise level and enhance the contrast.

- Hot spots – some of the basic units (usually called pixels, which term is derived from picture elements) may be faulty. This may mean that they always appear bright because of high noise levels, or they may impede the flow of electrons from other units. In the first case, the affected pixel can be corrected in a similar way to that for the cosmic ray spikes. In the second case, correction of the affected pixels may be possible by measuring the loss in the bad unit. Sometimes, however, the data in the pixels prior to the bad one are lost, and a blank line will have to be accepted in the image.

In general, image processing is undertaken for two purposes: firstly to correct known deficiencies in the image (e.g. flat fielding) and secondly to optimise the image to allow information to be retrieved by human beings (e.g. contrast stretching). There are many other techniques to image processing such as removal of the instrumental profile, use of false colours, removal of geometric distortion, smoothing, image combination, edge enhancement, Lagrange transforms, Fourier transforms, etc. In the second of the applications, image processing becomes much more of an art than a science, and which technique to apply and when, in order to obtain the information required, becomes a matter of judgement and trial-and-error. These other techniques are beyond the scope of this book, and the interested reader is referred to the specialist literature (Appendix 2).

Photography

The main imaging technique available on small telescopes, other than CCDs, remains photography. This has the advantages of being cheap, of providing a permanent image that is more-or-less independent of the observer and that is easy to archive, of allowing integration over time, of a high information density, and it is a familiar process to most people if only from holiday snaps. Its disadvantages are its very low quantum efficiency and hence slow speed, its non-linearity, and the complex, non-real-time processing that is required to obtain the image after the exposure.

Today's photographic emulsions are the result of over a century of development and therefore are quite sophisticated with a wide range of speeds, resolutions and spectral sensitivities. The basic detection mechanism in all types of photographic emulsion is, however, the same as that of the CCD – electron–hole pair production by an incident photon, but in silver bromide rather than silicon. Thereafter, however, the process differs from that of the CCD. The electron and hole are produced in a small crystal (grain) of the silver bromide, and are free to wander around within it under the effect of thermal motions. As with the CCD, they must be kept separate to avoid recombination. In the emulsion this is done chemically. The crystal structure of the silver bromide is deliberately distorted (strained) by the addition of a small number of chlorine atoms in solid solution (Figure 9.7). The strain in the crystal lattice where such an atom occurs results in a distortion of the electrical field, called a trap, which attracts the electron and repels the hole. The electron, immobilised at the trap, will neutralise a silver ion, leaving a silver atom in the crystal structure. The presence of the silver atom then adds to the effectiveness of the trap, so that the next electron produced by the absorption of a photon is more likely to be held there than at other traps. The second electron neutralises a second silver ion, further enhancing the effectiveness of the trap, and so on. Between five and 20 such silver atoms must accumulate before becoming stable against

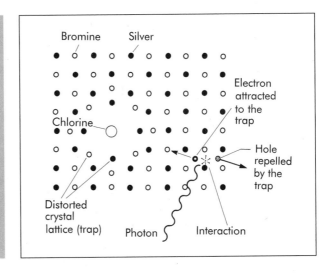

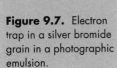

Figure 9.7. Electron trap in a silver bromide grain in a photographic emulsion.

dispersal by other processes, and such a cluster of atoms is then known as the latent image. The hole must be eliminated, just as with a CCD. In the emulsion it may be got rid of in two ways: either its random motions may take it to the surface of the grain, where it will react with the gelatine within which the grains are suspended, or two holes may combine to neutralise a halogen ion, which in turn will leak out of the grain to react with the gelatine.

Once the latent image has been produced, it must be processed. Processing is a chemical procedure whereby the latent image is converted into a visible image. It has at least two main stages: developing and fixing. Developers are reducing chemicals that convert silver bromide into silver. Normally the reaction rate for this conversion is very slow, but it is catalysed by the presence of silver atoms in the grains. The grains in the emulsion with latent images are thus converted completely into silver much more quickly than those grains without a latent image that are still pure silver bromide. Since a typical grain in the emulsion will contain some 10^{11} or so silver atoms, developing amplifies the latent image by a factor of about 10^9. A developer, however, will eventually convert all grains into silver, irrespective of the presence of a latent image or not. Hence the developing times must be strictly controlled. After developing, the silver grains forming the image are still surrounded by the undeveloped silver bromide grains. Fixing is a second chemical process that dissolves away the silver bromide. Fixers will also dissolve the silver grains given enough time, so again the process must only last for a strictly controlled reaction time. Practical details of photographic processing are to be found in many books (Appendix 2), and the manufacturer gives the specific requirements for an individual emulsion.

The final image, after processing, consists of the silver grains suspended in gelatine, the density of the grains being highest where the original light intensity was greatest. On viewing such an image, the silver grains, with a size in the region of 1 µm, absorb the light. The image therefore appears darkest where the original intensity was highest, i.e. it is a negative image. Most astronomical work takes place directly on the negatives; something that appears strange at first, but which quickly becomes familiar. If required, a photograph of the photograph may be made (usually called a print), to produce a positive image. The details of printing are again left to the specialist literature (Appendix 2).

Many astronomical photographs use black and white film, but colour film can also be used. Commercial colour films are usually optimised towards short exposures, and their colour balance may become incorrect on the long exposures characteristic of astronomical photography. This problem can be reduced by cooling the emulsion with dry ice (frozen carbon dioxide), or overcome completely by taking three separate black and white exposures through colour filters that are then combined in the print with appropriate exposures to give a correct colour rendition. Most of the colour images from larger telescope are obtained by this latter method.

p-i-n Photodiode

Of the many other types of light detectors available, only the p-i-n photodiode is likely to be used with small to medium-sized telescopes. This is a point detector rather than an imaging detector, and finds wide application in simple photometers (Chapter 11). It is a solid-state device, with one side made of p-type silicon, the other of n-type silicon, and with a thin layer of undoped or intrinsic silicon separating the two (hence the name). At the junction of two dissimilar materials a voltage usually develops (the Seebeck effect, for details see the specialist literature on condensed matter physics), and this phenomenon is also used in the thermocouple. The detection mechanism in the photodiode, like the CCD, is electron–hole pair production in silicon, with the electron and hole this time being separated by the Seebeck potential across the junction. The electron and hole produced by an interacting photon move across the junction and represent a flow of current in the device. That current may then be detected and its value provides a measure of the intensity of the illuminating radiation. The photodiode can be used in various modes, but all require only relatively simple electronics, which are suitable for construction by an averagely competent DIY enthusiast. Details of individual photodiodes and their associated circuits are supplied by the manufacturers.

Superconducting Tunnel Junction Detectors

A detector operating on quite different principles from those discussed above seems likely to replace the CCD for use on large telescopes in the next few years. It was originally developed for use as an X-ray detector, but is now finding application for infrared, visual and ultraviolet work. Not only does it have a high quantum efficiency from 100 nm to 2 μm, but the wavelength of each photon is measured as well. In other words, objects are not only detected, but their spectra (Chapter 12) are determined and with a resolution of a few nanometres.

The detector is called the superconducting tunnel junction (STJ or Josephson junction) and its detection mechanism is the disturbance of an electrical current within a superconductor. In a superconductor, linked electrons known as Cooper pairs carry the current. An absorbed photon can disrupt hundreds of such pairs, and the higher its energy (the shorter its wavelength) the more pairs are affected.

The superconductor used in an STJ is a metal such as niobium, tantalum or hafnium. It has to be cooled to less than 1 K (below –272 °C) in order to reduce noise. This requirement will therefore limit the use of the detector to major observatories, where advanced and sophisticated cryogenic equipment is available. Further details of STJs are beyond the scope of this book and the interested reader is referred to sources listed in Appendix 2.

Exercise

9.1. The following graph shows the intensity distribution of the pixels in the CCD image of an isolated galaxy (the size of the galaxy is significantly smaller than the area covered by the

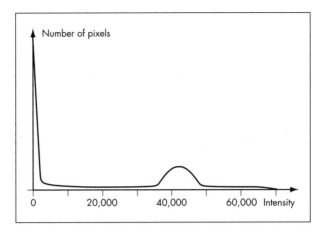

CCD). Suggest a suitable grey scaling (contrast stretch) to display the galaxy to best advantage on a monitor with 256 distinguishable grey levels. The CCD can detect a maximum intensity of 100 000.

Chapter 10

Data Processing

Introduction

Observing time on major telescopes is usually allocated to individual astronomers or groups of astronomers a few nights at a time, and at most three or four such sessions might normally be available over a year. This does not mean that the astronomers concerned can relax for the other 50 weeks, but it is a measure of the relative efforts required at a research level in obtaining the data in the first place compared with the processing then needed. Usually many multiples of the time spent observing have to be spent afterwards working on the data (photographs, CCD images, spectra, photometric measurements, astrometric measurements, etc.) in order to get it into the finally desired form.

Such an imbalance in the division of effort has been neither usual nor generally appropriate among observers using smaller instruments. Many such observers use their telescopes for the joy of gazing into the universe and, especially to begin with, are not concerned with dissecting out the gory details; secondly, there has perhaps seemed little that could be done with a small telescope by the way of astronomical research that would not be done first or better on the larger telescopes; and thirdly data processing often requires large amounts of computational effort which were unavailable outside the large research institutions. Photography, of course, requires the observer to spend time developing and printing his or her images, and sometimes this involves techniques such as dodging and unsharp masking that can be very time consuming. A century or more ago, most observations were obtained with what would now be classed as small telescopes, and were studied in depth, but in the recent past little effort has normally gone into processing the observations obtained with smaller telescopes.

That situation has in the last few years changed fundamentally. The advent of the CCD means that the owner of a 0.2 m telescope can undertake observations that only a decade ago would have required a 2 m telescope (Chapter 9), and for a quarter of the price of a 0.2 m Schmidt–Cassegrain telescope, that observer can purchase a computer with the memory and processing power of a mainframe computer of a few years ago. The 2 m telescope, of course, also uses a CCD, and so can make observations that would once have needed a 20 m telescope, but the universe is large enough to make a plethora of interesting and largely unstudied objects now available to all sizes of telescope. Original research is thus open to most astronomers who care to undertake it, and this means that data processing will have to become much more widely used than in the past.

Data processing breaks down roughly into two main subprocesses – data reduction and data analysis – and we consider each of these briefly below.

Data Reduction

This does not mean making your data smaller! It is the "mechanical" side of data processing, that is, the process or processes required to correct known faults or problems in data, and then to get the data into the most convenient form for the next stage, which is data analysis. Parts or all of data reduction can often be done automatically if the data are stored in the computer or available in computer-readable form. The exact stages involved in data processing vary with the observations and with the ultimate purpose for which they are to be used, but some examples will demonstrate the type of operation involved:

- Removal of cosmic ray spikes from CCD images
- Removal of the background, and the flat fielding of CCD images (Chapter 9)

- Contrast stretching, false colour representation, and other image processing techniques (Chapter 9)
- Determination of the response (characteristic curve) of a photographic emulsion and the conversion of the photographic density of the original image back into intensity
- Correction of geometrical distortions in the image
- Smoothing or other noise reduction procedures, such as adding together several images. Generally a signal must be at least as strong as the noise level in order for the object to be regarded as detected (a signal to noise ratio, or S/N, of 1). In practice most work requires S/Ns of 10–1000 or more. Images, especially those from radio telescopes may be processed in various ways to reduce the noise. The main such procedures for this are known as CLEAN and maximum entropy methods (MEM). However, the details of these are beyond the scope of this book and the interested reader is referred to sources in Appendix 2.
- Correction for expansion or contraction or other temperature-induced defects
- Reduction of the blurring effect (usually called the instrument profile or point spread function) of the telescope and other ancillary instruments used to obtain the data
- Removal of electrical mains supply interference or of other cyclic defects
- Conversion of intensity values into stellar magnitudes
- Calibration of the wavelength along a spectrum
- Correction for atmospheric absorption
- Correction for the effect of Earth's velocity on wavelengths (Doppler shift) and position in the sky (aberration)
- Correction for the effects of any filters used while obtaining the observations
- Calculation of means, standard deviations and standard errors of the mean for a set of measurements, etc.

Space is not available to consider all these processes in detail here, but the last mentioned is so fundamental that familiarity with it is essential to any observer intending to progress beyond the simplest of observations. Whenever you make measurements, you should always try to estimate or measure the errors/uncertainties in those measurements. The most direct way of determining the random errors is to repeat the measurements many times. The scatter in the results then gives an idea of the errors. That scatter is quantified by the standard deviation:

$$\sigma = \sqrt{\frac{\sum_{i=1}^{n}(X-x_i)^2}{n-1}} \qquad (10.1)$$

where σ is the standard deviation, n is the number of measurements, x_i is the ith measurement and X is the mean value (average) of the measurements.

The symbol Σ means that the quantity following it is summed over all possible values of x_i from $i = 1$ to $i = n$, i.e.

$$\sum_{i=1}^{n}(X-x_i)^2 = (X-x_1)^2 + (X-x_2)^2 + (X-x_3)^2 + \ldots + (X-x_n)^2. \qquad (10.2)$$

The standard deviation gives a measure of how far from the true value a single measurement is likely to lie; 68% of measurements should lie within 1σ of the true value, 95% of measurements within 2σ of the true value, and 99% of measurements within 2.5σ of the true value (see the discussion below, within the section about the correlation coefficient, on the use of the terms *"significant"* and *"highly significant"* in this context).

The standard error of the mean, S, is a measure of how far from the true value the mean or average of a number of measurements is likely to lie:

$$S = \frac{\sigma}{\sqrt{n}}. \qquad (10.3)$$

As with the standard deviation, the probability of a mean value being close to the true value is that 68% of mean values are likely to be within $1S$ of the true value, 95% within $2S$ of the true value, and 99% within $2.5S$ of the true value. Thus repeating measurements does not make any individual measurement likely to be more correct than any other, but the mean value of the measurements should (slowly) approach the true value. The standard deviation and the standard error of the mean are themselves uncertain. Unless you are using more than about a hundred measurements, both quantities can only be relied upon to about one significant figure.

If you cannot repeat the measurements, you should still try realistically to estimate the errors, for example from the known precision of your measuring instrument, and the degree of "fuzziness" of the point you are measuring, etc.

Numerical results should be presented in the form

$$X \pm S. \qquad (10.4)$$

Do not give the mean value to more significant places than are justified by the error estimate. Thus 12.5678 ± 0.0002 is acceptable, but 12.55678 ± 0.1345 should be given at the final stage (more figures can be retained during the calculation, and remember that only one significant figure is normally justified for the standard error of the mean) as 12.6 ± 0.1.

Data Analysis

Data analysis is the conversion of the data after reduction into an astronomically interesting form. Often it should more appropriately be called data synthesis, and it is where the astronomer usually makes his or her most significant contribution to the whole observing process. Data analysis varies even more than data reduction in the stages involved. At its simplest, it may be just the comparison of a new measurement of the magnitude of a star with one made previously to see if it has varied in brightness. At its most complex, it may involve the detailed computer modelling of a stellar or planetary atmosphere or interior, interstellar nebula, galaxy, cluster, etc. in order to determine rates of nucleosynthesis, element abundances, velocity, temperature or density structures, sources of energy, and so on, and it may need to include data of many different types, from many different sources, require complex theories, abstruse mathematics, and advanced concepts from physics and other sciences.

Data analysis thus depends both upon the nature of the observations and the purpose for which they are intended. It is not possible to give any sort of general guide to what may be involved. Some of the processes are discussed in Chapters 8, 11 and 12, but the observer will need to seek more specialist guidance than it is possible to give in an introductory book when he or she gets to the stage of undertaking original research.

There are numerous statistical procedures that can be applied within data analysis, and many of these require advanced knowledge of statistics and/or have only specialist applications. Three statistical techniques, however, are designed to answer basic questions that crop up very frequently in experimental work:

1 What equation best fits the data that I have obtained?

2 Do my measurements of quantity A show that it is related to quantity B?

3 Has the object that I am measuring changed between the two occasions when I studied it?

These three questions are addressed by the statistical techniques of linear regression, the correlation coefficient, and the Student t test. The practical use of these techniques is outlined below; their theoretical background, however, is left for the interested reader to pursue in more specialist books on statistics and data analysis.

Linear Regression

This technique is also known as linear least squares curve fitting. It finds the linear equation that is the "best" fit to a set of measurements of pairs of (possibly) related quantities, such as the periods and luminosities for Cepheid stars, sunspot numbers and terrestrial magnetic activity, radial velocities of galaxies and their distances, etc. That is to say, for a set of measurements of x and y, it finds values of a and b that give the "best" fit to the data in an equation of the type

$$y = ax + b. \qquad (10.5)$$

The technique has to be used with some caution. Firstly, the word "best" has been put here in inverted commas, because it is defined in an arbitrary manner. Linear regression produces the line for which the sum of all the squares of the distances of the measured points in the y direction from the line is minimised (Figure 10.1). It can therefore give excessive weightings to data points well away from the average. Other definitions of the difference between the line and the data points could be used, such as the modulus of the

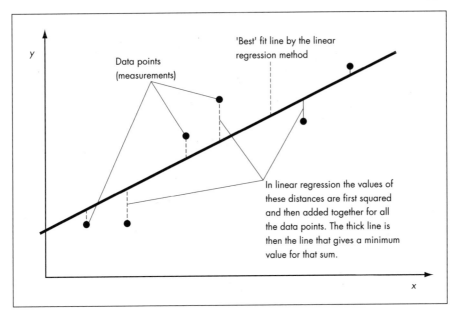

Figure 10.1. The least squares sum.

distance from the line or the logarithm of the distance, etc. However, these are mathematically more difficult to manipulate, and so the "least squares" solution is generally adopted.

The second problem with linear regression is that it will always give an answer, even when the quantities are connected in a non-linear fashion such as an exponential or sine relationship, or when there is no relationship at all. It has therefore to be used with care and with some physical insight into the processes being studied. The significance of the relationship determined by linear regression can also be assessed using the correlation coefficient (see below).

The third problem is that linear regression assumes that one set of data has a much higher level of precision than the other. This is often the case, for example in the light curve of a variable star, the times of the observations will be very much more accurately known than the magnitudes of the star. The formula then assigns the x variable to this high precision data set (the data can simply be relabelled if it is the y set that is more accurate). However, if both measurements have significant uncertainties, then linear regression may not give the best fitting curve.

With the above cautions, the use of the linear regression method just requires the use of two slightly forbidding, but nonetheless straightforward equations.

Using the same notation as for Equation (10.1), the values of a and b in Equation (10.5) are given by:

$$a = \frac{\sum_{i=1}^{n}(X-x_i)(Y-y_i)}{\sum_{i=1}^{n}(X-x_i)^2} \qquad (10.6)$$

$$b = Y - aX. \qquad (10.7)$$

As an example, we might well expect on theoretical grounds that the luminosity, l (in units of the solar luminosity, 4×10^{26} W), of a globular cluster would be related to the number of stars, n, in the cluster. Thus for the following set of measurements of the absolute magnitudes and masses of some globular clusters:

Cluster	l	n
M3	302 000	210 000
M5	209 000	60 000
M4	21 000	60 000
M13	159 000	300 000
M92	159 000	140 000
M22	48 000	7000 000
M15	331 000	6000 000

we have (with the luminosity substituting for x, and the number of stars for y)

$$a = \frac{\sum_{i=1}^{n}(L-l_i)(N-n_i)}{\sum_{i=1}^{n}(L-l_i)^2} = \frac{5.58 \times 10^{10}}{8.15 \times 10^{10}} = 0.685 \qquad (10.8)$$

and

$$b = N - aL = 1\,970\,000 - 0.685 \times 175\,600 = 1\,850\,000 \qquad (10.9)$$

and so we have the linear formula (taking reasonable account of the number of significant figures in the original data) relating the mass of a globular cluster to its absolute magnitude:

$$n = 0.69l + 1\,900\,000 \qquad (10.10)$$

However before triumphantly using this formula, see the example given below for the correlation coefficient.

Correlation Coefficient

Since linear regression produces a formula whether the two quantities are related or not, it is usually necessary to check how significant the correlation of the measurements may be, before using the formula for other purposes. Calculation of the correlation coefficient performs this task.

However, the question that the use of the correlation coefficient *actually* answers is "What is the probability that the measurements are *not* related?" A highly probable result from the use of the correlation coefficient therefore implies that the quantities measured are *not* related, while a low probability result implies that they *are* related. This approach to assessing probabilities is quite common within statistics (see also Student's *t* test below), but it can cause confusion when first encountered. However, the test can never tell you that two quantities are definitely related; it only gives the probability of that being the case.

By convention, a correlation is called *significant* if its probability is 5% (i.e. there is a 95% chance that the quantities are related) and *highly significant* if its probability is 1% (i.e. there is a 99% chance that the quantities are related). A research paper would not normally be accepted for publication if a claimed correlation was poorer than *significant* (the result of the test was a probability higher than 5%, i.e. the chance that the quantities are related was less than 95%). This conventional use of the terms *significant* and *highly significant* also applies in other circumstances such as Student's *t* test (below), standard deviation and the standard error of the mean (above).

The calculation of the value of the correlation coefficient uses a formula (Equation 10.11) similar to that involved in linear regression. (Note that if you are using both techniques, some quantities are duplicated between the formulae, and so do not have to be calculated twice.) But the interpretation of the result requires the use of a graph (Fig 10.2) or a set of statistical tables, and introduces a new quantity – the *number of degrees of freedom*. This is a statistical concept whose theoretical background is beyond the scope of this book. The practical realisation of the number of degrees of freedom for the correlation coefficient however is simple. It is just one less than the total number of pairs of data points.

The value of the correlation coefficient, r, is determined via the formula

$$r = \frac{\sum_{i=1}^{n}(X - x_i)(Y - y_i)}{\sqrt{\sum_{i=1}^{n}(X - x_i)^2 \sum_{i=1}^{n}(Y - y_i)^2}}. \qquad (10.11)$$

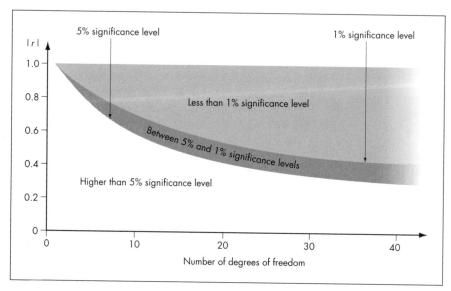

Figure 10.2. The correlation coefficient.

It is then interpreted from Fig. 10.2 (r has a value of +1 when the two quantities are perfectly correlated; of –1, when the two quantities are perfectly anti-correlated; and a value of 0 when they are uncorrelated).

Looking at the same set of data that was used as the example for linear regression, we thus have

$$r = \frac{\sum_{i=1}^{n}(L-l_i)(N-n_i)}{\sqrt{\sum_{i=1}^{n}(L-l_i)^2 \sum_{i=1}^{n}(N-n_i)^2}} = \frac{5.5806 \times 10^{10}}{\sqrt{8.149 \times 10^{10} \times 5.8072 \times 10^{13}}}$$

$$= \frac{5.5806 \times 10^{10}}{2.1752 \times 10^{12}} = 0.0257 \quad (10.12)$$

Since there are seven pairs of measurements, the number of degrees of freedom is six. From the graph in Fig. 10.2 we may therefore see that the result of the correlation coefficient test is a probability very much higher than 5%. Since this is the probability that the quantities being measured are not related, we may state (on the basis of this data at least) that there is no correlation between a globular cluster's luminosity and the number of stars that it contains.

Thus the second problem mentioned in connection with linear regression, that it will always give a solution, even when the data are uncorrelated, is illustrated. We may also see this from the plot in Figure 10.3 of the original data set and the linear regression equation.

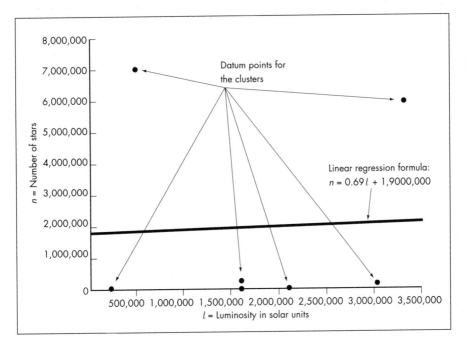

y-axis: n = Number of stars
x-axis: l = Luminosity in solar units

Datum points for the clusters

Linear regression formula:
$n = 0.69\,l + 1,9000,000$

Figure 10.3. The data on globular clusters' luminosities and numbers of stars, and the equation resulting from the application of the linear regression formulae.

Student's *t* Test[6]

This test tries to answer the question "Do two sets of measurements of the same object or quantity at different times (or wavelengths, positions, polarisations, etc.) show that it has changed?"

Like the correlation coefficient, the test is based on the assumption that the two sets of data *do not* differ from each other, and determines the probability that any observed difference is due to random errors only. The lower the probability given by the test, therefore, the higher the probability that the two sets of measurements *do* show that the object has changed. Thus a result evaluated at 1% by Student's *t* test gives a 99% chance of the measured quantity having changed.

Theoretically any value of the probability between 0 and 1 (0% to 100%) can be found from applying Student's *t* test, though in practice it will never be possible to state with *certainty* that a quantity has or has

---

[6] This is not a simplified test suitable for students that is to be superseded by a better test for higher levels of work. The name comes from the statistician who devised the test, William Gosset. He wrote popular articles under the pen-name Student.

not changed. As with the correlation coefficient, the values 5% and 1%, are chosen arbitrarily as decision points and sets of measurements evaluated at these two levels are termed *significant* and *highly significant*, respectively.

Student's t parameter is found from the equation

$$t = \left| \frac{X_A - X_B}{\sqrt{\left(\frac{\sigma_A^2}{N_A} + \frac{\sigma_B^2}{N_B} \right)}} \right| \tag{10.13}$$

where A and B are the two sets of measurements, X is the average value for the set, σ its standard deviation and N the number of measurements in the set.

For Student's t test the number of degrees of freedom is the number of measurements in each set minus 2 (i.e. $N_A + N_B - 2$). The significance of the result of the test may then be read off the graph in Figure 10.4.

An example of an application of Student's t test would be to determine if a star's brightness had changed between two observing sessions:

First set of measurements (A)

8.310 8.296 8.322 8.284 8.285 8.321 8.309
8.302 8.324 8.306 8.294 8.289 8.295 8.347 8.294

Second set of measurements (B)

8.173 8.313 8.257 8.258 8.299 8.312 8.293
8.284 8.315 8.289 8.271 8.263

Thus

$$N_A = 15 \qquad X_A = 8.305 \qquad \sigma_A = 0.017$$
$$N_B = 12 \qquad X_B = 8.285 \qquad \sigma_B = 0.021$$

giving

$$t = 2.67.$$

The number of degrees of freedom is 25. So from the graph in Figure 10.4, we may see that the result of Student's t test is a probability slightly less than 1%. Therefore the result is *highly significant* and there is greater than 99% chance that the star is a variable.

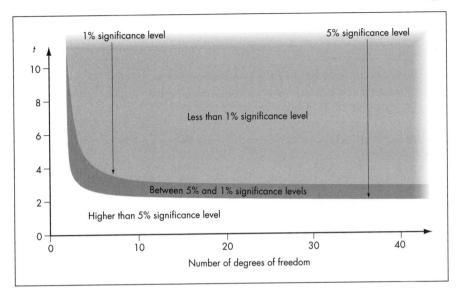

205

Figure 10.4.
Student's *t* test.

Exercises

10.1 Calculate the mean, standard deviation, and standard error of the mean of the following set of measurements of the Sun's angular diameter in seconds of arc:

959.6 959.3 959.5 959.8 959.75 959.0 960.1
959.7 959.65 959.4

10.2 Express the following numbers correctly:

12.556 78 ± 0.1345
3.59 ± 0.01
5002 ± 103
0.001 2345 ± 0.000 6789
$3.09 \times 10^6 \pm 10^5$
$3.318 \times 10^{16} \pm 2.7 \times 10^{16}$

10.3 Find a linear relationship between the apparent magnitudes and numbers of asteroids in each one-magnitude range.

Apparent magnitude range	Number of asteroids
4.5 ± 0.5	2
5.5 ± 0.5	1
6.5 ± 0.5	6
7.5 ± 0.5	25
8.5 ± 0.5	80
9.5 ± 0.5	160
10.5 ± 0.5	320
11.5 ± 0.5	800

Hence predict the number of asteroids in the range 15^m to 16^m. How realistic is this answer likely to be?

10.4 A connection is suspected between the number of astronomy students at an observing session and the temperature. Is there a real correlation?

Number of students	20	10	15	21	20	5	25	25	25	28
Temperature (°C)	5	–7	–8	0	–5	–9	9	5	15	10

10.5 The separation of a close double star has been measured on two occasions. Has the separation changed over time?

First set of measurements in seconds of arc (A)

1.29   1.33   1.34   1.35   1.27

Second set of measurements in seconds of arc (B)

1.27   1.30   1.26   1.28   1.25   1.27

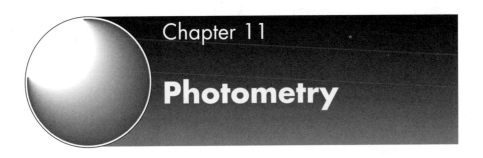

Photometry

Introduction

Photometry is the process of determining intensity or magnitude of or within the original object. It most commonly involves measurements of point sources or near-point sources like stars and planets, but it can also involve measuring the integrated intensities or magnitudes of more extended objects. Any of the detectors discussed in Chapter 9 can be used for photometry. The CCD and p-i-n photodiode are, however, the most straightforward because they have linear responses, and their outputs can be converted directly into magnitudes using Equation (8.5). The eye and the photographic emulsion are non-linear detectors and their responses are therefore more complex to convert into magnitudes. The visual estimation of magnitude having already been discussed in Chapter 8, we shall consider CCD and photographic photometry further below.

CCD Photometry

This is straightforward since the CCD response is linear. Scintillation, however, will normally spread the images, even of stars, over several pixels. The actual intensity must therefore be found by adding the intensities of those pixels together. Fortunately most manufacturers of astronomical CCDs supply software that enables this to be done, and in some cases to give the stars' magnitudes directly. For this, however, there will need to be some standard stars (stars with known magnitudes) within the image as well as the star(s) of interest.

Filters will normally need to be used to restrict the range of wavelengths being detected (see the section on UBV photometry below). If you wish to compare your magnitudes with those measured elsewhere, then the CCD–filter combination will need to give the same overall response as the standard detector–filter combinations. Since the wavelength sensitivity of the CCD differs markedly from that of the eye, photomultiplier and p-i-n photodiode, filters sold for use with those detectors will not give the correct response when used with a CCD. Most manufacturers of astronomical CCDs, however, will also supply suitable filters for use with them.

The dynamic range of CCDs is large (×100 000 or more – equivalent to 12.5 stellar magnitudes). However, it is possible to saturate the image, and the magnitudes then obtained will be incorrect. If the magnitudes of a very bright star and a very faint star are both needed, it may be necessary to take two images, one short enough not to saturate the image of the bright star, and the other long enough to give a good image of the faint star. Some CCDs have extra electrodes that drain off electrons as pixels approach saturation. This process is known as anti-blooming. The images will not then be saturated, but the effect is to make the CCD response non-linear at high intensities. You will need to check with the manufacturer of your CCD to find out if your device operates in this way. If it does, then you will only be able to use it for photometry on images that are well away from being saturated.

Photographic Photometry

Photometric measurements from a photographic image are generally made using a microdensitometer or from the diameters of the stellar images. The latter approach requires the least equipment: just a medium-power magnifier with a finely divided scale and an illuminated screen on which to rest the negative. The

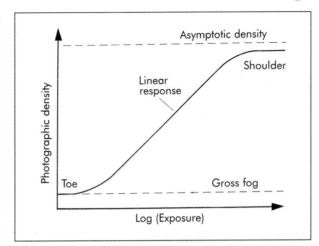

Figure 11.1.
Characteristic curve of a photographic emulsion (schematic).

diameter of a stellar image varies with its brightness because of the intrinsic background noise, called the gross fog (Figure 11.1), always present on a photographic image. The image of a star fades from a maximum at its centre to zero over a distance of a second of arc or two because of the intrinsic spread (Figure 8.4), aberrations, and the effect of atmospheric turbulence. The point at which that image rises above the gross fog therefore varies with brightness (Figure 11.2), and so the observed image diameter also varies with brightness. If three or four or more stars with known magnitudes are on the image, their diameters can be used to plot a diameter–magnitude graph, and this can then be used to obtain the magnitudes of the other stars on the image.

The microdensitometer is a device that shines a light through the emulsion and accurately measures the

Figure 11.2.
Transects of photographic stellar images in the presence of fog. Note that the actual diameters of the two images are the same, but the visible image diameters differ because of the proportion of the image that exceeds the fog level.

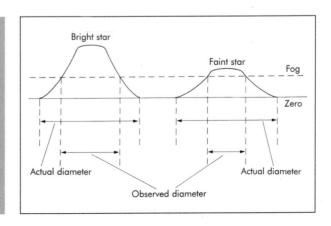

transmitted intensity. Microdensitometers are more usually used for measurements of non-point sources and spectra (Chapter 12), but for images that are not saturated (overexposed) they can be used to measure stellar brightnesses. The transmitted intensity is related to the photographic density, D, by

$$D = \log_{10} I_{in} - \log_{10} I_{out} \qquad (11.1)$$

where I_{in} is the intensity of the radiation illuminating the emulsion, I_{out} is the intensity of the radiation emerging from the emulsion, and the photographic density is related in turn to the original intensity in the image by the characteristic curve of the emulsion (Figure 11.1). In order to find the original intensities (and magnitudes), the characteristic curve has therefore to be known. The characteristic curve is very sensitive to emulsion manufacturing conditions, exposure type, temperature, age, developing processes, etc. Thus a calibration exposure, a photograph of a series of sources of known relative intensities, has to be made under as similar conditions as possible (same emulsion batch, processed simultaneously, similar spectral region, etc.) for every exposure, though sometimes a few exposures on identical emulsions from the same batch can use the same calibration. The calibration exposure is then measured with the microdensitometer, and used to plot the characteristic curve.

Absolute Magnitude

The magnitudes so far considered are all as the objects *appear* in the sky. They are therefore termed apparent magnitudes, and are denoted by the lower case m. The relationship between energy and apparent magnitude was discussed in Chapter 8 (Equation 8.5). Clearly, apparent magnitude tells us nothing about the actual energy or luminosity of the object. Thus the Moon (Table 8.2) has an apparent magnitude of –12.7, and Betelgeuse one of –0.73, but the Moon is not actually 60 000 times brighter than Betelgeuse – we just see it so because the Moon is so much closer.

To provide a quantity that is related to the actual luminosity of the object, we use the absolute magnitude (denoted by the upper case M). The absolute magnitude is defined as the apparent magnitude that the

Table 11.1. Absolute magnitudes

Object	M
Sun	+4.8
Full Moon	+32
Venus (maximum)	+29
Jupiter (maximum)	+26
Sirius A	+1.4
Betelgeuse	–6
Polaris	–4.6
Type 1a supernova	–19
Faintest white dwarfs	+16

object would have if its distance were 10 parsecs, and so differences in absolute magnitudes reflect real differences in the brightnesses of the objects. The absolute and apparent magnitudes are related by

$$M = m + 5 - 5\log_{10} D \qquad (11.2)$$

where D is the distance of the star in parsecs.

The absolute magnitudes of some objects are listed in Table 11.1, from which we may see that the true relationship between the Moon and Betelgeuse is that the latter is 10^{15} times brighter than the former! However, see also the section on bolometric magnitude, below.

Wavelength Dependence

Stars and many other astronomical objects have spectra that are approximately black-body in overall shape (Figure 11.3). A hotter star will therefore be relatively brighter in the blue parts of the spectrum compared with a cooler star. Conversely, the latter will be relatively brighter at longer wavelengths. The value obtained for the magnitude therefore depends upon the wavelength at which it is measured. Magnitudes estimated by eye are called visual magnitudes (m_v or M_v), and correspond roughly to the relative energies of the objects at 510 nm. Magnitudes obtained from blue-sensitive photographs are called photographic magnitudes (m_p or M_p), and correspond to the energies around 440 nm. Magnitudes obtained from orthochromatic photographs, which mimic the spectral sensitiv-

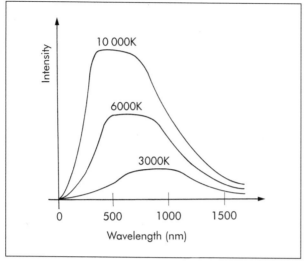

Figure 11.3.
Black-body curves.

ity of the eye, are called photovisual (m_{pv} or M_{pv}). CCDs and other detectors such as photomultipliers have spectral sensitivities that differ from each other, and from those of the eye and the various types of photographic emulsion. It is therefore usual to use filters to isolate the wave band over which a magnitude is to be measured. There are innumerable such filter systems currently in use by astronomers, many of which are designed for highly specific and restricted purposes, such as monitoring the strength of a particular spectrum line, and have little merit outside that application. The most commonly used system of filters is, however, in very widespread use, and is called the *UBV* system. It is the only one that we shall consider here.

UBV System

The commonest of the photometric filter systems in the visual is called the *UBV* system. It uses three filters, centred on 365 nm (*Ultraviolet*), 440 nm (*Blue*) and 550 nm (*Visual*), each with a bandwidth of about 100 nm. The magnitudes obtained through the filters are designated *U*, *B* and *V*. The *V* magnitude thus corresponds roughly to that which would be estimated by eye. The *B* magnitude corresponds roughly to that obtained using blue-sensitive photographic emulsion. The *U* magnitude does not have a visual or photographic analogue, and is badly affected by atmospheric absorption,

Table 11.2. Bolometric correction (BC)	
Spectral class	BC
O5	4.0
B5	1.5
A5	0.12
F5	0
G5	0.07
K5	0.6
M5	2.3

so is of little use except from high altitude sites. One of the main reasons for undertaking photometry is that it can provide a simple method of determining the temperature of a star. In the UBV system, the temperature is obtained from the difference between the B and V magnitudes (usually called the $B-V$ colour index) using the semi-empirical relationship

$$T = \frac{8540}{(B-V)+0.865} K. \qquad (11.3)$$

Bolometric Magnitude

The true luminosity of a star is given by the absolute bolometric magnitude, M_{Bol}. The bolometric magnitude takes account of the energy emitted at all wavelengths from gamma rays to radio waves. The bolometric magnitude may be found from the star's absolute magnitude in the blue-green region (V or M_V) and its spectral class (Chapter 12). The bolometric correction (BC – Table 11.2) is the amount that must be subtracted[7] from the visual absolute magnitude to give the bolometric absolute magnitude, i.e.

$$M_{Bol} = M_V - BC. \qquad (11.4)$$

The total energy emitted by the star (i.e. its luminosity) is then

$$\text{Luminosity} = 3 \times 10^{28} \times 10^{-2 \times M_{Bol}/5} \text{ W} \qquad (11.5)$$

---

[7] In some sources the bolometric correction may be listed as a negative number that must then be *added* to the visual magnitude to give the bolometric magnitude.

Spatial Information

Much of photometry is concerned only with the brightnesses of point or small sources. Sometimes, however, the variation of brightness with position is important. Thus in spectroscopy (Chapter 12) the variation of intensity with wavelength is needed, in the image of an extended object the precise variation of intensity across it may be required, or its integrated magnitude may be wanted. Such information may be obtained by repeated or scanned observations with a point detector, but is generally more easily found from an image. With a CCD image, each pixel is essentially a point photometer and the magnitude of its charge is directly related to the intensity at that point. A photographic image, however, needs to be scanned and the photographic density converted to intensity using the characteristic curve, for every point in the image. This is the real purpose of the microdensitometer, rather than the measurement of stellar magnitudes (see above). Although designs for different machines vary in their details, the photograph may normally be placed on a glass-surfaced platform that may be moved uniformly in one or two dimensions, so covering the whole image of the object. The output from the microdensitometer is then usually fed into a computer for further processing.

Photometers

Magnitudes may be estimated by eye or found from CCD or photographic images as just discussed, but for the highest precision work, instruments specifically designed to obtain brightness measurements, known as photometers, must be used. These vary widely in their details, but most designs are based upon the basic photometer (Figure 11.4). This has a small entrance aperture to reduce background noise, a finding and guiding system, a means of interchanging filters rapidly, and a detector. The latter, for visual work, is usually a photomultiplier or a p-i-n photodiode. Often the detector needs to be cooled to reduce thermal noise. A Fabry lens may also be used; this focuses the telescope objective on to the detector and provides a uniform image that does not move over the detector's surface with the scintillation motion of the source or with tracking

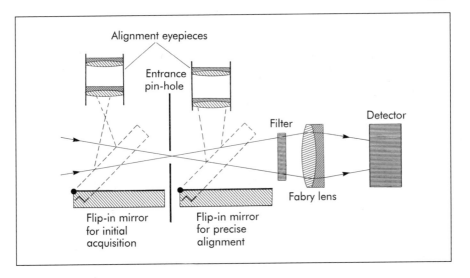

Alignment eyepieces

Entrance pin-hole

Filter

Detector

Fabry lens

Flip-in mirror for initial acquisition

Flip-in mirror for precise alignment

Figure 11.4. Basic photometer.

errors. Small photometers based upon p-i-n photodiodes are produced commercially, but a CCD can be used in their place much of the time.

Observing Techniques

Photometry using a photometer requires precise observing techniques if it is to give accurate results. A standard star for comparison is needed and this should be close to the star of interest in the sky, of similar brightness, and of similar spectral type (Chapter 12). The observing sequence is then to measure first the comparison star, then the star of interest, and then the comparison star again to ensure that the observing conditions or the instrument settings, etc. have not changed during the time taken for the observations. For each filter that is to be used, first the sky background is measured, then the star, then the background again, and this sequence is repeated four or five times. This procedure is also to ensure that conditions are not changing and to enable a measure of the uncertainties (Chapter 10) to be obtained. The output to a chart recorder of such an observing run is shown schematically in Figure 11.5. Magnitudes can be obtained from the data using Equation (8.5) and then colour indices, temperatures, absolute magnitudes, etc. calculated.

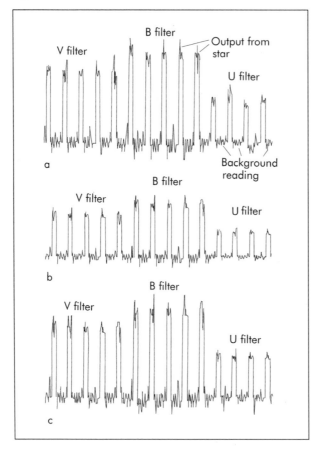

Figure 11.5.
Schematic chart recorder output from an observing run with a basic photometer: (a) first set of readings for the comparison star; (b) set of readings for the star of interest; (c) second set of readings for the comparison star.

Although magnitudes can be obtained from single CCD or photographic images, it is better to use several such images, so that uncertainties, changing sky conditions, etc. can similarly be estimated.

Exercises

11.1 A standard star, of apparent V magnitude 7.31, gives an output from a photometer of 0.362 mV. What are the apparent V magnitudes of two nearby stars for which the outputs from the same photometer are 4.079 mV and 209 μV?

11.2 From what distance could we just detect Jupiter, if observing in the visible with the best of present-day techniques, but from well outside the solar system (ignore contrast problems with the Sun)? Suggest whether or not we should be able to detect comparable planets around other stars by this method.

11.3 Determine the absolute bolometric magnitude and the total energy emitted by Betelgeuse using the information in Tables 11.1 and 11.2, given also that its spectral class is M2 I. The Sun emits 4×10^{26} W, so how much brighter is Betelgeuse than the Sun?

Chapter 12

Spectroscopy

Introduction

Spectroscopy is the study of the way that the brightness of an object varies with wavelength. *UBV* photometry gives some information of this type (Chapter 11); the STJ detector (Chapter 9) also has an intrinsic spectral resolution (see below) of a few tens of nanometres in the visual region. But, by common consent, spectroscopy normally has a spectral resolution of 1% or better, photometry a spectral resolution of 1% or worse. Although it is the most fruitful technique available to astronomers, capable of yielding information on temperatures, compositions, luminosities, pressures, magnetic fields, levels of excitation and ionisation, surface structure, line of sight velocities, turbulent velocities, rotational velocities, expansion/contraction, binarity, and, less directly, distances, masses and ages, spectroscopy has generally found little favour among users of smaller telescopes. This is almost certainly because of the long exposures normally required, even on large telescopes, in order to obtain a spectrum. With a CCD detector, however, spectroscopy becomes eminently possible for telescopes of 0.2 or 0.3 m aperture and above and small spectroscopes are now available commercially. This chapter is therefore written more in the hope of encouraging wider interest in the topic among observers with such instruments than as a guide to current practice.

Spectroscopes

Spectral resolution and dispersion are the two primary criteria governing the performance of spectroscopes. They are defined by

$$\text{Spectral resolution} = R = \frac{\lambda}{\Delta\lambda} \qquad (12.1)$$

$$\text{Dispersion} = \frac{\delta\lambda}{\delta x} \qquad (12.2)$$

where λ is the operating wavelength, $\Delta\lambda$ is the smallest wavelength interval that can be distinguished and $\delta\lambda$ is the change in wavelength over a distance δx along the spectrum.

Spectral resolutions used in astronomy range from 10 to 100 000 ($\Delta\lambda$ = 50 to 0.005 nm for visual work), and dispersions from 200 nm mm^{-1} to 0.01 nm mm^{-1}. The lower resolutions and poorer dispersions mostly arise, however, in searches and survey work, and so minimum useful values are more normally about 300–500 for spectral resolution and 10–20 nm mm^{-1} for dispersion. Such levels of performance are achievable by quite a simple spectroscope using a small prism or diffraction grating. There is room here only to outline what is involved in designing a spectroscope, and the interested reader is referred to more specialist books (Appendix 2) for further information.

The basic spectroscope contains six major elements (Figure 12.1): an entrance slit to give a pure spectrum and to limit background noise, a widener, a collimator,

Figure 12.1. Basic spectroscope (shown with a diffraction grating as the dispersing element, although this can be replaced by a prism).

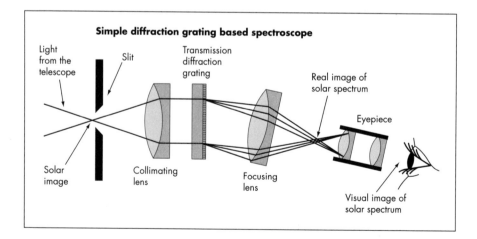

Simple diffraction grating based spectroscope

Light from the telescope

Slit

Transmission diffraction grating

Real image of solar spectrum

Solar image

Collimating lens

Focusing lens

Eyepiece

Visual image of solar spectrum

a dispersing element, a focusing element and a detector. The final spectrum is composed of overlapping monochromatic images of the entrance slit, and this is why features in the spectrum are normally seen as lines. The slit should thus be as narrow as possible, without excluding too much light from the object being observed, and its sides should be accurately parallel to each other. If guiding is to be undertaken, then this is usually accomplished by having the outer faces of the slit jaws polished and then observing the overspill of light from the object at the sides of the slit. The widener is to broaden the spectrum at right angles to its length, otherwise stellar spectra would be too narrow to use. An oscillating parallel-sided block is often used, or the observer can use the telescope drives to trail the image up and down the slit during the exposure. The collimator and focusing element are conventional optical components and can be either lenses or mirrors. The dispersing element separates the light into its component wavelengths. It can be either a small prism or a diffraction grating. In the latter case it should be blazed (its individual apertures angled to direct the light into the desired spectral order) or it will waste most of the light. The detector can be a photographic emulsion or a CCD, but in the former case the observer will need to be prepared for very long exposures. As a very rough guide, a widened spectrum with a dispersion of 10 nm mm^{-1} for a magnitude 4 star obtained on a 0.3 m telescope would require an exposure of between half an hour and four hours with a photographic detector, and about 1% to 10% of that with a CCD detector.

The spectroscopes used on large telescopes can be very complex and cost an appreciable fraction of the value of the telescope itself. This is because so much information is obtainable from a spectrum. The spectroscope is thus designed to operate at a range of dispersions, resolutions and wavelength ranges, and to be as efficient as possible. In the last decade or so, the efficiency has been improved dramatically by obtaining the spectra of many objects simultaneously. The spectroscope is fed by a number of fibre optic cables ("light pipes"). The other ends of the cables are positioned, under computer control, at the focal plane of the telescope in such a way as to coincide with the images of the objects of interest. For example, the two-degree field (2df) instrument on the 3.9 m Anglo-Australian telescope can obtain up to 400 spectra in a single exposure.

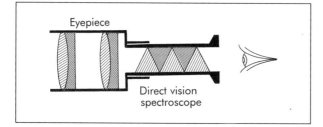

Figure 12.2. Use of a direct vision spectroscope for visual spectroscopy.

For visual work, a much simpler approach can be used which will enable the main spectrum lines to be seen, as well as the differences between spectral types (see below) and stellar and nebular spectra. A small direct vision spectroscope of the type sold to chemists for element identification in flame spectra can be attached to the end of a conventional eyepiece (Figure 12.2), and will enable low-dispersion spectra of the brighter stars to be seen directly.

Spectroscopy

Reference has already been made to the wide range of information derivable from spectra. An indication of how that information is obtained is summarised below, but the interested reader will need to consult more specialised books (Appendix 2) before being able to undertake serious work in any of these areas.

Spectral Type

One of the quickest ways of extracting information is via the spectral type of a star. This is a classification system based upon eye estimates of the features visible on medium-dispersion spectrograms. With experience, an observer can find a great deal of information about the star for a small investment of his or her time and effort.

Early in the history of stellar spectroscopy, it was seen that stellar spectra showed certain recurring themes that enabled them to be grouped together. One of these early classification systems was based upon the relative intensities of the hydrogen lines with respect to the other lines in the spectrum. Spectra with the strongest hydrogen lines were group A, those with slightly weaker lines, group B, slightly weaker still, group C, and so on. The differences between spectra were initially attributed to evolution, with the spectra changing in the sense of decreasing hydrogen line intensity and of increasing complexity, as the star became older. The least complex spectra, types A, B, etc. were thus thought to come from young (or early) stars, while the more complex spectra came from older (or later) stars.

Stellar spectra do, of course, change as the star ages, but not in any such simple-minded manner. The main underlying reason for the differing appearances of stellar spectra is variation in their surface temperatures. A more useful stellar spectral classification would thus be based upon a temperature sequence of the spectra, and that is how the present system is arranged. Unfortunately, the change from the earlier classification method to that of the present day was accomplished by rearranging and adapting the older system, not by starting afresh. The present spectral classification system is thus now rather untidy and unnecessarily difficult to use. It seems unlikely to be rationalised, however, so the student must become familiar with it as it stands.

The modern system of classification was originally codified in Harvard's Henry Draper star catalogue, and so is sometimes known as the Harvard classification system. The Harvard system was then further developed by Morgan, Keenan and Kellman at Yerkes, and is now more commonly known as the MKK system. There are 13 groups of spectra, of which seven, labelled by the letters,

O B A F G K M

form the core. The O type stars are the hottest normal stars (50 000 K), and the M type the coolest (3000 K). A useful mnemonic for the order of the classes is

Oh Be A Fine Girl/Guy Kiss Me.

Each of the major classes is subdivided into ten, with Arabic numerals denoting the subdivisions. The Sun, for example, is of spectral class G2, while Betelgeuse is M2 and Sirius A2. The first three groups are often called early type stars, while the last two are called late type stars. This, however, is now just a convenient convention, and no longer has the evolutionary significance once attributed to it.

Recently the original decimal subdivision of the core classes has been adapted to provide a smoother variation with temperature by adding and deleting some of the classes. The full range of spectral types in current use is thus:

Core	Subdivisions								
O	4	5	6	7	8	9	9.5		
B	0	0.5	1	2	3	5	7	8	9.5
A	0	2	3	5	7				
F	0	2	3	5	7	8	9		
G	0	2	5	8					
K	0	2	3	4	5				
M	0	1	2	3	4	7	8		

The missing numbers though are likely still to be encountered, especially in older sources.

The other six letters

R N S W P Q

are much more rarely used and denote unusual groups of stars such as the carbon stars (R and N, sometimes also called C-type stars), and the Wolf–Rayet stars (W), etc.

The main features of the core spectral types are:

O   Few lines present; those actually visible are mostly due to highly ionised silicon and nitrogen, etc. Hydrogen Balmer and ionised helium lines visible.

B   Balmer lines strengthening, neutral helium and lower stages of ionisation of silicon and nitrogen, etc. now producing lines. Towards the lower temperature end of the group, the neutral helium lines disappear.

A   Balmer lines peak in intensity at A0; singly ionised calcium lines appear. Thereafter the Balmer lines weaken and numerous lines due to singly ionised metals start to appear.

F   Lines in the spectra become more numerous, Balmer and ionised metal lines continue to weaken, neutral metal lines strengthen.

G   Ca II, H and K lines peak in their intensities, Balmer lines continue to weaken, neutral metal lines continue to increase in intensity.

K   The Balmer lines are still just visible, many lines due to neutral metals are now present, and a few molecular bands (TiO) appear.

M   The TiO bands now dominate the spectrum.

The precise determination of the spectral class of a star from its spectrum is via the intensity ratios of pairs of lines that happen to be especially sensitive to temperature or luminosity. The required line pairs vary with spectral class, some only being used to distinguish one or two subclasses, others being of wider use. The student is referred to specialist works on the topic (Appendix 2) for further details of the techniques of spectral classification. Once the spectral class of a star is determined, then its temperature and many other properties may be found from tables (Appendix 2).

Luminosity Class

The widths of spectral lines increase as the gas pressure in the regions producing them increases. Now the masses of normal stars vary by only about a factor of 1000, while their diameters vary by a factor of about 10,000, and their volumes by a factor of 10^{12}. The physically largest stars must therefore have much lower densities than the physically smallest stars, and hence the larger stars have the lower surface pressures. Thus the spectral lines originating from large, and hence very luminous, stars generally have smaller widths than those from smaller stars. The overall effect of differing pressures is to change the intensity ratios between some pairs of lines in the spectra from stars that have identical temperature-based spectral classes, but differing luminosities. The luminosity class is added as a Roman numeral after the temperature spectral class. Classes I to IV are the giant stars (supergiants, bright giants, giants and subgiants respectively). The majority of stars, including the Sun, fall into class V, and are called main sequence or dwarf stars. Classes VI and VII are the subdwarfs and white dwarfs. A more complete classification for the earlier examples is therefore: the Sun G2 V, Betelgeuse M2 I and Sirius Al V.

Radial Velocity

After spectral classification, the velocity of an object along the line of sight, usually known as its radial velocity, is the commonest parameter to be obtained from a spectrum. For this purpose a comparison spectrum is usually required. This is an emission-line spec-

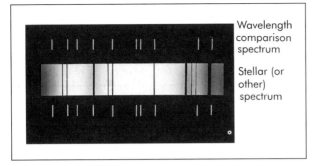

Wavelength comparison spectrum

Stellar (or other) spectrum

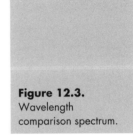

Figure 12.3.
Wavelength comparison spectrum.

trum from an artificial source. The comparison spectrum is placed either side of the main spectrum (Figure 12.3). The wavelengths of the lines in the comparison spectrum are known, and so those of lines in the main spectrum may be found by interpolation. These will normally be different from the rest wavelengths of those same lines because of the Doppler shift and so the object's velocity can be found from the Doppler formula:

$$\text{Radial velocity} = v = \frac{c\Delta v}{v} = \frac{c\Delta \lambda}{\lambda} \qquad (12.3)$$

where v is the velocity of the object along the line of sight, c is the velocity of light (Chapter 7), $\Delta\lambda$ and Δv are the wavelength or frequency shifts: $\Delta\lambda = \lambda_O - \lambda_L$, $\Delta v = v_L - v_O$, where subscript "O" denotes an observed value, and subscript "L" denotes a laboratory (unshifted) value.

The convention is used that the radial velocity is positive when directed away from the Earth, and negative when directed towards the Earth.

Spectrophotometry

The greatest return of information from a spectrum comes from the detailed study of spectrum line strengths and of their shapes (profiles) – this is known as spectrophotometry. The process usually requires extensive computer modelling of the region producing the spectrum. The models are used to produce predicted spectra, and these are then compared with the observations. The model is adapted until as close a fit as possible to the observations is obtained, and the properties of the observed object inferred from the best-fit model. The process can become extremely complex, and any further discussion has to be left to more advanced texts.

Exercise

12.1 Calculate the line-of-sight velocity with respect to the Earth of a star when the Balmer H-α line in its spectrum is observed to be at a wavelength of 655.2 nm. (Balmer H-α line rest wavelength: 656.2868 nm; speed of light in a vacuum: 2.998×10^8 m s^{-1}.)

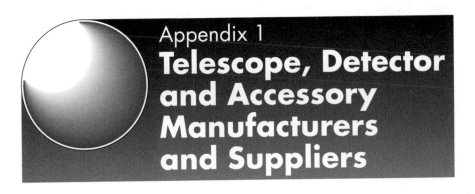

Appendix 1

Telescope, Detector and Accessory Manufacturers and Suppliers

Note that inclusion of a supplier or manufacturer in the list below does not constitute a recommendation or endorsement by the author or publishers of the supplier's products. Readers are advised to obtain full information and competitive quotes before making any purchases. An indication of the type of product is shown in brackets – telescopes (T), detectors (D) and accessories (A).

Abel Express, (A)
KDKK Industrial Complex,
Building 2, 100 Rosslyn Rd,
Carnegie,
PA 15106,
USA

AE Optics Ltd, (T)
Vega Court,
East Drive,
Caldecote,
Cambridge CB3 7NZ,
United Kingdom

Ash Manufacturing Company, (Domes)
Plainfield,
IL 60544,
USA

Astronomical Innovations, (A)
PO Box 14853,
Lenexa,
KS 66285,
USA

Astro-physics Inc., (T)
11250 Forest Hills Rd,
Rockford,
IL 6115,
USA

Broadhurst Clarkson and Fuller, (T, D, A)
Telescope House,
63 Farringdon Rd,
London EClM 3JB,
United Kingdom

Carl Zeiss Jena GmbH, (T)
Postfach 125,
Tatzendpromenade 1A,
Jena, 0-6900,
Germany

Celestron Telescopes, (T)
PO Box 3578,
Torrance,
CA 90503,
USA

Coronado, (A)
1674 South Research Loop, Suite 436,
Tucson,
AZ 85710
USA

Dark Star Telescopes, (T)
6 Pinewood Drive,
Ashley Heath,
Market Drayton,
Salop,
United Kingdom

Day Star Filters, (A)
PO Box 5110,
Diamond Bar,
CA 91765,
USA

David Hinds Ltd, (T)
Unit 34,
The Silk Mill,
Brook St,
Tring, HP23 5EF,
United Kingdom

Eclipse Ltd, (A)
Belle Etoile,
Rue du Hamel,
Castel,
Guernsey, GY5 7QJ,
United Kingdom

Helios Telescopes, (T, A)
Optical Vision Ltd,
Unit 2b, Woolpit Business Park,
Woolpit,
Bury St Edmunds,
Suffolk, IP30 9UP,
United Kingdom

Konus Corp., (T)
8359 NW Street, 68,
Miami,
FL 33166
USA

Lumicon, (T)
2111 Research Dr., 5,
Livermore,
CA 94550,
USA

Meade Instrument Co., (T, A)
16542 Millikan Avenue,
Irvine,
CA 93714,
USA

Orion Optics, (T)
Unit 12,
Quakers Coppice,
Crewe Gates Industrial Estate,
Crewe,
Cheshire, CW1 1FA,
United Kingdom

Questar Corp., (T)
Route 202,
Box 59,
New Hope,
PA 18938,
USA

Roger W. Tuthill Inc., (T, D, A)
Box 1086,
11 Tanglewood Lane,
Mountainside,
NJ 07092-0086,
USA

Santa Barbara Instrument Group, (D, A)
147-A Castilian Drive,
Santa Barbara,
CA 93117,
USA

Scope City, (T)
71 Bold St,
Liverpool, LI 4EZ,
United Kingdom

Scopes Direct, (T)
Unit 22,
Third Avenue,
Crewe,
Cheshire, CW1 6XU,
United Kingdom

Starlight Xpress Ltd., (D)
Foxley Green Farm,
Ascot Rd.,
Holyport,
Berks, SL6 3LA,
United Kingdom

Swift Instruments Inc., (T)
952 Dorchester Avenue,
Boston,
MA 02125,
USA

TAL, (T)
Maly Zlatoustinsky 8,
Suite 2,
Moscow 101000,
Russia

Thousand Oaks Optical, (A)
Box 4813,
Thousand Oaks,
CA 91359,
USA

Torus Optical, (T)
67 Bon-Aire,
Iowa City,
IA 52240,
USA

Unitron, (T)
170 Wilbur Place,
PO Box 469,
Bohemia,
NY 11716,
USA

Vixen Co. Ltd, (T)
5–17 Higashitokorozawa,
Tokorozawa,
Saitama 359-0021,
Japan

Bibliography

Journals

Only the major and relatively widely available journals are listed. There are numerous more specialised research-level journals available in academic libraries.

Popular

Astronomy
Astronomy Now
Ciel et Espace
Journal of the British Astronomical Association
New Scientist
Practical Astronomy
Publications of the Astronomical Society of the Pacific
Scientific American
Sky and Telescope

Research

Astronomical Journal
Astronomy and Astrophysics
Astrophysical Journal
Monthly Notices of the Royal Astronomical Association
Nature
Science

Star and Other Catalogues, Atlases and Reference Books

Astronomical Almanac, HMSO/US Government Printing Office (published annually).

Astrophysical Quantities, CW Allen, Athlone Press, 1973.

Atlas of Representative Stellar Spectra, Y Yamashita, K Nariai and Y Norimoto, University of Tokyo Press, 1977.

Cambridge Deep-Sky Album, J Newton and P Teece, Cambridge University Press, 1983.

Cambridge Encyclopedia of the Sun, KR Lang, Cambridge University Press, 2001.

Deep Sky Observer's Year, G Privett and P Parsons, Springer-Verlag, 2001.

Deep Sky Observing, SR Coe, Springer-Verlag, 2000.

Encyclopedia of Astronomy and Astrophysics, edited by P Murdin, *Nature* and IoP Publishing, 2001.

Encyclopedia of Planetary Sciences, edited by JH Shirley and RW Fairbridge, Kluwer Academic Publishers, 2000.

Field Guide to the Deep Sky Objects, M Inglis, Springer-Verlag, 2001.

Handbook of the British Astronomical Association, British Astronomical Association (published annually).

Illustrated Dictionary of Practical Astronomy, CR Kitchin, Springer-Verlag, 2002.

Messier's Nebulae and Star Clusters, KG Jones, Cambridge University Press, 1991.

Norton's 2000.0, edited by I Ridpath, Longman, 1998.

Observer's Sky Atlas, E Karkoschka, Springer-Verlag, 1999.

Observer's Year, P Moore, Springer-Verlag, 1998.

Observing Handbook and Catalogue of Deep Sky Objects, C Luginbuhl and B Skiff, Cambridge University Press, 1990.

Observing the Caldwell Objects, D Ratledge, Springer-Verlag, 2000.

Photographic Atlas of the Stars, H Arnold, P Doherty and P Moore, IoP Publishing, 1997.

Photo-Guide to the Constellations, CR Kitchin, Springer-Verlag, 1997.

Planetary Nebulae: A Practical Guide and Handbook for Amateur Astronomers, SJ Hynes, Willmann-Bell, 1991.

Seeing Stars, CR Kitchin and R Forrest, Springer-Verlag, 1997.

Sky Atlas 2000.0, W Tirion, Sky Publishing Corporation, 2000.

Sky Catalogue 2000, Volumes 1 and 2, A Hirshfield and RW Sinnott, Cambridge University Press, 1992.

Yearbook of Astronomy, Macmillan, published annually.

Practical Astronomy Books

Amateur Telescope Making, SF Tonkin, Springer-Verlag, 1999.

Analysis of Starlight; 150 Years of Astronomical Spectroscopy, JB Hearnshaw, Cambridge University Press, 1987.

Art and Science of CCD Astronomy, D Ratledge, Springer-Verlag, 1997.

Astronomical Equipment for Amateurs, M Mobberley, Springer-Verlag, 1999.

Astronomical Spectroscopy, CR Kitchin, Adam Hilger, 1995.

Astronomy on the Personal Computer, O Montenbruck and T Pfleger, Springer-Verlag, 1991.

Astronomy with Small Telescopes, SF Tonkin, Springer-Verlag, 2001.

Astrophysical Techniques, CR Kitchin, Adam Hilger, 1998.

Building and Using an Astronomical Observatory, P Doherty, Stevens, 1986.

Challenges of Astronomy, W Schlosser, T Schmidt-Kaler and EF Malone, Springer-Verlag, 1991.

Choosing and Using a Schmidt–Cassegrain Telescope, R Mollise, Springer-Verlag, 2001.

Compendium of Practical Astronomy, GD Roth, Springer-Verlag, 1993.

Computer Processing of Remotely-Sensed Images: An Introduction, PW Mather, John Wiley, 1987.

Data Analysis in Astronomy, V Di Gesu, L Scarsi and R Buccheri, Plenum Press, 1992.

Exercises in Practical Astronomy using Photographs, MT Bruck, Adam Hilger, 1990.

Getting the Measure of Stars, WA Cooper and EN Walker, Adam Hilger, 1989.

Introduction to Experimental Astronomy, RB Culver, WH Freeman, 1984.

Manual of Advanced Celestial Photography, BD Wallis and RW Provin, Cambridge University Press, 1988

Modern Amateur Astronomer, edited by P Moore, Springer-Verlag, 1995.

Observational Astronomy, DS Birney, Cambridge University Press, 1991.

Observing Meteors, Comets, Supernovae, N Bone, Springer-Verlag, 1999.

Observing the Moon, P Wlasuk, Springer-Verlag, 2000.

Observing the Sun, PO Taylor, Cambridge University Press, 1991.

Practical Astronomer, CA Ronan, Pan, 1981.

Practical Astronomy: A User Friendly Handbook for Skywatchers, HR Mills, Albion, 1993.

Practical Astronomy with your Calculator, PD Smith, Cambridge University Press, 1981.

Practical Astronomy: A User Friendly Handbook for Skywatchers, HR Mills, Albion, 1993.

Seeing the Sky: 100 Projects, Activities and Explorations in Astronomy, F Schaaf, John Wiley, 1990.

Software and Data for Practical Astronomers, D Ratledge, Springer-Verlag, 1999.

Solar Observing Techniques, CR Kitchin, Springer-Verlag, 2001.

Solar System: A Practical Guide, D Reidy and K Wallace, Allen and Unwin, 1991.

Star Gazing through Binoculars: A Complete Guide to Binocular Astronomy, S Mensing, TAB, 1986.

Star Hopping: Your Visa to the Universe, RA Garfinkle, Cambridge University Press, 1993.

Using the Meade ETX, M. Weasner, Springer-Verlag, 2002.

Workbook for Astronomy, J Waxman, Cambridge University Press, 1984.

Introductory Books

AstroFAQs, SF Tonkin, Springer-Verlag, 2000.

Astronomy: A Self-Teaching Guide, DL Moche, John Wiley, 1993.

Astronomy on the Personal Computer, O Montenbruck, T Pfleger, Springer-Verlag, 2000.

Astronomy: The Evolving Universe, M Zeilik, John Wiley, 1994.

Astronomy: Principles and Practice, AB Roy and D Clark, Adam Hilger, 1988.

Astronomy through Space and Time, S Engelbrektson, WC Brown, 1994.

Eyes on the Universe, P Moore, Springer-Verlag, 1997.

Introductory Astronomy, K Halliday, John Wiley, 1999.

Introductory Astronomy and Astrophysics, M Zeilik, SA Gregory and EvP Smith, Saunders, 1992.

Unfolding our Universe, I. Nicolson, Cambridge University Press, 1999.

Universe, RA Freedman and WJ Kaufmann III, WH Freeman, 2001.

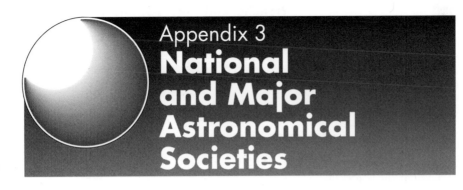

For details of local astronomical societies see *International Directory of Astronomical Associations and Societies*, published annually by the Centre de Données de Strasbourg, Université de Strasbourg.

Agrupación Astronáutica Española,
Rosellón 134,
E-08036 Barcelona,
Spain

American Association for the Advancement of Science,
1333 II Street NW,
Washington,
DC 2005,
USA

American Association of Variable Star Observers,
25 Birch St,
Cambridge,
MA 02138,
USA

American Astronomical Society,
2000 Florida Avenue NW,
Suite 3000,
Washington,
DC 20009,
USA

Association Française d'Astronomie,
Observatoire de Montsouris,
17 Rue Emile-Deutsch-de-la-Meurthe,
F-75014 Paris,
France

Association des Groupes d'Astronomes Amateurs,
4545 Avenue Pierre-de-Coubertin,
Casier Postal 1000, Succ M.,
Montreal,
QC H1V 3R2,
Canada

Association Nationale Science Techniques Jeunesse, Section
Astronomique,
Palais de la Découverte,
Avenue Franklin Roosevelt,
F-75008 Paris,
France

Association of Lunar and Planetary Observers,
PO Box 16131,
San Francisco,
CA 94116,
USA

Astronomical-Geodetical Society of Russia,
24 Sadovaja-Kudrinskaya Ul.,
SU-103101 Moscow,
Russia

Astronomical League,
6235 Omie Circle,
Pensacola,
FL 32504,
USA

Astronomical Society of Australia,
School of Physics,
University of Sydney,
Sydney,
NSW 2006,
Australia

Astronomical Society of the Pacific,
1290 24th Avenue,
San Francisco,
CA 94122
USA

Astronomical Society of Southern Africa,
Southern African Astronomical Observatory,
PO Box 9,
Observatory 7935,
South Africa

Astronomisk Selskab,
Observatoriet,
Øster Volgade 3,
DK-1350 Copenhagen K,
Denmark.

British Astronomical Association,
Burlington House,
Piccadilly,
London, W1V 9AG
United Kingdom

British Interplanetary Society,
27/29 South Lambeth Rd,
London SW8 1SZ,
United Kingdom

Canadian Astronomical Society,
Dominion Astrophysical Observatory,
5071 W. Saanich Rd,
Victoria,
BC V8X 4M6
Canada

Committee on Space Research (COSPAR),
51 Bd de Montmorency,
F-75016 Paris,
France

Earthwatch,
680 Mount Auburn St,
Box 403,
Watertown,
MA 02272,
USA

Federation of Astronomical Societies,
1 Tal-y-Bont Rd.,
Ely,
Cardiff, CF5 5EU,
Wales

International Astronomical Union,
61, Avenue de l'Observatoire,
F-75014 Paris,
France

Junior Astronomical Society,
10 Swanwick Walk,
Tadley,
Basingstoke,
Hampshire RG26 6JZ,
United Kingdom

National Space Society,
West Wing Suite 203,
600 Maryland Avenue SW,
Washington,
DC 20024,
USA

Nederlandse Astronomenclub,
Netherlands Foundation for Radio Astronomy,
Postbus2,
NL-7990 AA Dwingeloo,
Netherlands

Nederlandse Vereniging voor Weer-en Sterrenkunde,
Nachtegaalstrat 82 bis,
NL-3581 AN Utrecht,
Netherlands

Nippon Temmon Gakkai,
Tokyo Tenmondai,
2-21-1 Mitaka,
Tokyo 181,
Japan

Royal Astronomical Society,
Burlington House,
Piccadilly,
London, W1V 0NL,
United Kingdom

Royal Astronomical Society of Canada,
136 Dupont St,
Toronto, 0NT M5R 1V2,
Canada

Royal Astronomical Society of New Zealand,
PO Box 3181,
Wellington,
New Zealand

Schweizerische Astronomische Gesellschaft,
Hirtenhoffstrasse 9,
CH-6005 Lucerne,
Switzerland

Società Astronomica Italiana,
Osservatorio Astrofisico di Arcetri,
Largo E. Fermi 5,
I-50125 Florence,
Italy

Société d'Astronomie Populaire,
1 Avenue Camille Flammarion,
F-31500 Toulouse,
France

Société Astronomique de France,
3 Rue Beethoven,
F-75016 Paris,
France

Société Royale Belge d'Astronomie, de Météorologie, et de
Physique du Globe,
Observatoire Royale de Belgique,
Avenue Circulaire 3,
B-1180 Brussels,
Belgium

Stichting De Koepel,
Nachtegaalstrat 82 bis,
NL-3581 AN Utrecht,
Netherlands

Svenska Astronomiska Sallskapet,
Stockholms Observatorium,
S133 00 Saltsjöbaden,
Sweden

Vercinigung der Sternfreunde e.V.,
Volkssternwarte,
Anzingerstrasse 1,
D-8000 Munich,
Germany

Zentral Kommission Astronomie und Raumfahrt,
Postfach 34,
DDR-1030 Berlin,
Germany

Appendix 4

Constellations

Constellation	Abbreviation	Constellation	Abbreviation
Andromeda	And	Leo	Leo
Antlia	Ant	Leo Minor	LMi
Apus	Aps	Lepus	Lep
Aquarius	Aqr	Libra	Lib
Aquila	Aql	Lupus	Lup
Ara	Ara	Lynx	Lyn
Aries	Ari	Lyra	Lyr
Auriga	Aur	Mensa	Men
Boötes	Boo	Microscopium	Mic
Caelum	Cae	Monoceros	Mon
Camelopardalis	Cam	Musca	Mus
Cancer	Cnc	Norma	Nor
Canes Venatici	CVn	Octans	Oct
Canis Major	CMa	Ophiuchus	Oph
Canis Minor	CMi	Orion	Ori
Capricornus	Cap	Pavo	Pav
Carina	Car	Pegasus	Peg
Cassiopeia	Cas	Perseus	Per
Centaurus	Cen	Phoenix	Phe
Cepheus	Cep	Pictor	Pic
Cetus	Cet	Pisces	Psc
Chamaeleon	Cha	Piscis Austrinus	PsA
Circinus	Cir	Puppis	Pup
Columba	Col	Pyxis	Pyx
Coma Berenices	Com	Reticulum	Ret
Corona Australis	CrA	Sagitta	Sge
Corona Borealis	CrB	Sagittarius	Sgr
Corvus	Crv	Scorpius	Sco
Crater	Crt	Sculptor	Scl
Crux	Cru	Scutum	Sct

Cygnus	Cyg	Serpens	Ser
Delphinus	Del	Sextans	Sex
Dorado	Dor	Taurus	Tau
Draco	Dra	Telescopium	Tel
Equuleus	Equ	Triangulum	Tri
Eridanus	Eri	Triangulum Australe	TrA
Fornax	For	Tucana	Tuc
Gemini	Gem	Ursa Major	UMa
Grus	Gru	Ursa Minor	UMi
Hercules	Her	Vela	Vel
Horologium	Hor	Virgo	Vir
Hydra	Hya	Volans	Vol
Hydrus	Hyi	Vulpecula	Vul
Indus	Ind		
Lacerta	Lac		

2.1 (a) ×112; (b) 0.61″; (c) 800

2.2 (a) ×42; (b) ×390

2.3 4.8×10^{-7} radians (= 0.099″)

4.1 2h 53m 9.5s
1h 4m 0s
6h 11m 50s
19h 26m 16s

4.2 46° 31' 15″
221° 30' 0″
272° 22' 0″
318° 18' 15″

4.3 6h 31m 4s

4.4 4h 42m 55s

4.5 23h 13m 35s

4.6 Sirius: 16h 28m 26s
Betelgeuse: 17h 18m 25s
Neither star would be visible

4.7 Altitude = +70° 20' 51″; azimuth = 107° 07' 53″ west

5.1 50° 16' 51″ east

5.2 ⩾ 82° 37'

5.3 6h 45m 13s
−16° 40' 31″

5.4 Venus: 1.597 years
Jupiter: 1.092 years
Pluto: 1.004 years

5.5 (a) No: lunar eclipses can only occur at full Moon.

(b) Yes: the Moon's orbit rotates in space once every 18.7 years, and so a solar eclipse can occur sometime anywhere around the Earth's orbit.

8.1 Sixth magnitude star: 9.87×10^{-11} W m^{-2}

9.1 The Galaxy extends from intensities of about 36 000 to 48 000. One (of many) suitable grey scalings would thus be:

CCD intensity	Monitor level
0–35 999	0
36 000–36 049	1
36 050–36 099	2
36 100–36 149	3
36 150–36 199	4
and so on up to	
48 600–48 649	253
48 650–48 699	254
48 700–100 000	255

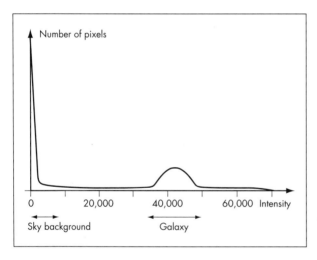

10.1 Mean = 959.58
$\sigma = 0.302$
$S = 0.0955$
Thus the final answer should be 959.6 ± 0.1

10.2 12.6 ± 0.1
3.59 ± 0.01
5000 ± 100
0.0012 ± 0.0007 or 0.001 ± 0.0007
$3.1 \times 10^6 \pm 10^5$
$3 \times 10^{16} \pm 3 \times 10^{16}$

10.3 Number of asteroids = 91.64 m − 558.9. (Number of asteroids = 90 m − 560 is more in line with the number of significant figures in the original data.) Hence the number of asteroids in the range 15^m to 16^m (m = 15.5) is 835. This predicted number is likely to be far too small since the original data show a much steeper than linear relationship (probably exponential) in the number of asteroids with decreasing brightness.

10.4 r = 0.85. With nine degrees of freedom, we find from Figure 10.4 that the significance level << 1%. The correlation of student numbers with temperature is therefore *highly significant*.

10.5 t = 2.59. The number of degrees of freedom is nine. So, from the graph in Figure 10.4, we may see that the result of Student's t test is a probability in the region between 5% and 1% (actually just slightly less than 5%). Therefore, the result is *significant* and there is greater than 95% chance that the stars' separation has changed with time.

11.1 Magnitudes: 4.67, 7.91

11.2 27 parsecs (for Jupiter to have an apparent magnitude of +28 − see Table 8.2); thus Jovian-sized planets could be detectable out to a few tens of parsecs.

11.3 For Betelgeuse, M_V = −6, and BC = 1.8 (by linear interpolation), so from Equation (11.4),

$$M_{Bol} = -7.8$$

So from Equation (11.5)

$$\text{Luminosity} = 2.3 \times 10^{31} \text{ W.}$$

Betelgeuse is thus 58 000 times brighter than the Sun.

12.1 496 km s^{-1} towards the Earth.

SI and Other Units

SI Prefixes

Prefix	Multiplier	Symbol
atto	10^{-18}	a
femto	10^{-15}	f
pico	10^{-12}	p
nano	10^{-9}	n
micro	10^{-6}	μ
milli	10^{-3}	m
centi	10^{-2}	c (not recommended)
deci	10^{-1}	d (not recommended)
deca	10^{1}	da (not recommended)
hecto	10^{2}	h (not recommended)
kilo	10^{3}	k
mega	10^{6}	M
giga	10^{9}	G
tera	10^{12}	T
peta	10^{15}	P
exa	10^{18}	E

SI Units

Physical quantity	Unit	Symbol
angle	radian	rad
capacitance	farad	F (s^4 A^2 m^{-2} kg^{-1})
electric charge	coulomb	C (A s)
electric current	ampere	A
electrical resistance	ohm	Ω (m^2 kg s^{-3} A^{-2})
energy	joule	J (m^2 kg s^{-2})
force	newton	N (kg m s^{-1})
frequency	hertz	Hz (s^{-1})
length	metre	m
luminous intensity	candela	cd
magnetic flux density	tesla	T (kg s^{-2} A^{-1})
mass	kilogram	kg
power	watt	W (m^2 kg s^{-3})
pressure	pascal	Pa (kg m^{-1} s^{-2})
solid angle	steradian	sr
temperature	kelvin	K
time	second	s
voltage	volt	V (m^2 kg s^{-3} A^{-1})

Other Units in Common Use in Astronomy

Unit	Symbol	Equivalent
Ångstrom	Å	10^{-10} m
astronomical unit	AU	$1.495\ 978\ 70 \times 10^{11}$ m
atmosphere	atm	$1.013\ 25 \times 10^5$ Pa
bar	bar	10^5 Pa
dyne	dyn	10^{-5} N
electron volt	eV	1.6022×10^{-19} J
erg	erg	10^{-7} J
gauss	G	10^{-4} T
jansky	jy	10^{-26} W m^{-2} Hz^{-1}
light year	ly	9.4605×10^{15} m
micron	μ, μm	10^{-6} m
parsec	pc	3.0857×10^{16} m
solar luminosity	$L_\odot$	3.8478×10^{26} W
solar mass	$M_\odot$	1.9891×10^{30} kg
solar radius	$R_\odot$	6.960×10^8 m

Greek Alphabet

Letter	Lower case	Upper case
Alpha	α	A
Beta	β	B
Gamma	γ	Γ
Delta	δ	Δ
Epsilon	ε	E
Zeta	ζ	Z
Eta	η	H
Theta	θ	Θ
Iota	ι	I
Kappa	κ	K
Lambda	λ	Λ
Mu	μ	M
Nu	ν	N
Xi	ξ	Ξ
Omicron	o	O
Pi	π	Π
Rho	ρ	P
Sigma	σ	Σ
Tau	τ	T
Upsilon	υ	Y
Phi	ϕ	Φ
Chi	χ	X
Psi	ψ	Ψ
Omega	ω	Ω

Index

Pages that mark the start of a chapter or major section on the topic are underlined.

aberration: chromatic, 6, 7, 8, 9, 14, 20, 48
aberration: optical, 5, 13, 14, 19, 48
aberration: spherical , 6, 7, 8, 9, 12, 19, 48, 49
aberration: stellar, 114, 115, 116
absolute / apparent magnitude relationship, 211
absolute magnitude, 166, 210, 211, 213
absorption: atmospheric, 23, 142, 143, 195
achromatic doublet: see achromatic lens,
achromatic lens, 12, 13, 14, 49
achromatism: condition for, 49
active supports, 22, 41, 55
active telescope mounting, 54, 55
aerial telescope, 8
Airy disk, 32, 34
aligning an equatorial telescope mounting , 130, 131, 132
Almagest, 109
alt-az mounting, 18, 55, 56, 57, 91, 126, 129
altitude, 80, 83
analemma, 86
angular measure: hours, minutes and seconds, 83
annual motions, 103, 105, 106
annular solar eclipse, 118, 120
Antarctic, 59, 105
anti-blooming: charge coupled device, 208
apastron, 121
aperture synthesis, 25, 27, 29, 54
aperture synthesis: filled aperture, 29, 54
aperture synthesis: unfilled aperture, 29, 30
apex , 116
aphelion, 121
apochromatic lens, 13, 14
apocynthion, 121
apogee, 118, 121
apojove, 121
apparent / absolute magnitude relationship, 211
apparent magnitude, 166, 167, 210, 211
appulse, 99, 118
apsis, 121

Arctic, 105
Arecibo telescope, 27
artificial guide star, 24
asterism, 150, 171
asteroid nomenclature, 163
astigmatism, 12, 19, 49, 50, 51
astrology, 79, 106
Astronomical Almanac, 88, 89, 108, 148, 158
astronomical refractor, 5, 15
atmospheric absorption, 23, 25, 142, 143, 195
atmospheric compensation, 23, 34
atmospheric refraction , 105, 134
atmospheric turbulence, 147
autoguider, 134
autumnal equinox, 82, 87, 104
averted vision, 38, 172, 177, 179
azimuth, 80, 83

Bacon, Roger, 4
ball and claw telescope mounting, 135
Barlow lens, 47
barn door mounting, 57
barrel distortion, 51, 52
Bayer, Johan, 163
Bayer system of stellar nomenclature, 163, 164
beat frequency, 26
bell metal, 11
binary star, 169
binoculars, 68, 69
binoculars: gyro-stabilised, 69
black body spectrum, 211, 212
blank: mirror, 39, 63
B magnitude, 212
bolometer, 26
bolometric correction, 213
bolometric magnitude, 213

calculation of local sidereal time, 89
Caldwell catalogue, 173
camera: digital , 184
Canary islands, 58
cardinal points, 77
Cassegrain telescope, 11, 18, 19, 22, 25, 68
Cassegrain, Guillaume, 11
Cassini. Giovanni, 9

Cassini's division, 9
catadioptric telescope, 19
catoptric telescope, 19
CCD: see charge coupled device,
celestial latitude, 93
celestial longitude, 93
celestial sphere, 75, 76, 101
Čerenkov radiation, 140
characteristic curve of a photographic emulsion, 195, 209, 210
charge coupled device, 24, 180, 181, 182, 183, 184, 194, 207, 221
charge coupled device: anti-blooming, 208
charge coupled device: cooling, 184
charge coupled device: dynamic range, 208
charge coupled device: efficiency, 184
charge coupled device: fluorescent coating, 184
charge coupled device images: cosmic ray spikes, 185, 194
charge coupled device images: hot spots, 186
charge coupled device: operating principle, 180, 181
charge coupled device: photometry, 207, 208
charge coupled device: read out, 184
charge coupled device: rear illumination, 184
charge coupled device: spectral response, 184
charge coupled device: thinned, 184
chart: finder, 149
chromatic aberration, 6, 7, 8, 9, 14, 20, 48
circle of least confusion, 50, 51
circumpolar object, 102, 103, 105
civil time, 86, 87, 102
CLEAN, 195
coelostat, 161
collimation, 66, 67
colour images: photography, 189
colour index, 213
coloured stars, 168
coma, 12, 19, 49, 50
comet filter, 48
comet nomenclature, 163

comparison star, 215
compass points, 77
compensation: atmospheric, 23
computer modelling, 197
condition for achromatism, 49
cone cells: retina, 31, 37, 177, 179
conjunction, 116, 117
constellations, 79, 243
contrast stretching, 185, 195
converging lens, 13
Cooper pairs of electrons, 190
Copernicus, Nicolaus, 5, 7, 110
correcting lens, 17, 20
correcting mirror, 24
correlation coefficient, 198, 200, 201, 202
cosine rule: Euclidean , 95
cosine rule: spherical , 97
cosmic ray spikes on charge coupled device images, 185, 194
Coudé telescope system, 18
cross wire eyepiece, 41, 131, 134

dark adaptation, 32, 37, 178
dark signal subtraction, 185, 194
data analysis, 194, 197, 198
data processing, 193, 194
data reduction, 194
data synthesis, 197
daytime observing, 173
declination, 56, 82, 83, 90, 91, 102, 126, 132
declination axis, 56, 126
Dec: see declination,
degree of freedom, 201, 202, 204, 205
detection principle of a photographic emulsion, 187
determination of the field of view: eyepiece, 150
developing a photographic emulsion, 188
dew cap, 146
dewing-up, 146, 148
diffraction, 32, 33, 34
diffraction grating, 220, 221
diffraction limit, 23, 43, 157, 170
Digges, Leonard, 4
digital camera, 184
dioptric telescope, 19
dipole: half-wave, 26
direct motion, 108, 109, 110
direct vision spectroscope, 222
dispersion, 13, 220
distances of the stars, 114
distortion, 49, 50
distortion: barrel, 51, 52
distortion: pin-cushion, 51, 52
diurnal motion, 101
diverging lens, 13
Dobsonian mounting, 67, 68, 129, 130
Dobsonian telescope, 67, 68
Dollond, John, 12, 13
dome, 14
double stars, 168, 169, 170
Dreyer, Johann, 173
dynamic range: charge coupled device, 208

Earl of Rosse: see Parsons, William,
earliest evening, 104
earliest morning, 104
Earth rotation synthesis: see aperture synthesis,
Earth's orbit, 80
Earth's orbital motion, 81, 86, 103, 114, 115
Earth's rotation, 81, 91, 126
eclipse, 117
eclipse: annular, 118, 120
eclipse: limits on, 120
eclipse: lunar, 119, 121, 152
eclipse: partial, 118, 120
eclipse: solar, 118, 120
eclipse: total, 118, 120
ecliptic, 78, 104
electromagnetic radiation, 139
electromagnetic radiation: interactions with matter, 142
electromagnetic radiation: polarisation, 141, 143
electromagnetic radiation: range of wavelengths, 142, 143
electron traps in a photographic emulsion, 187
electron volt, 140
electron-hole pairs, 181, 187, 189
elongation, 116, 117
e-m radiation: see electromagnetic radiation,
English telescope mounting , 126, 128
ephemeris, 108
equation of time, 86
equator, 76
equatorial mounting, 18, 55, 56, 91, 126
equatorial mounting: alignment, 130, 131, 132
equatorial platform, 68, 129
equinox, 82, 104
Erfle eyepiece, 44, 45
errors in data, 195, 196, 197
ESO: see European Southern Observatory,
estimated uncertainties, 197
European Southern Observatory, 22
exit pupil, 34, 35
extended image, 35
extended object: surface brightness, 36, 37
eye, 177, 178
eye: dark adapted, 32, 37, 178
eye estimates of magnitudes, 168
eyepiece, 5, 8, 13, 15, 34, 41, 65, 147, 148
eyepiece: cross wire, 41, 131, 134
eyepiece: determination of the field of view, 150
eyepiece: Erfle, 44, 45
eyepiece: field of view, 35, 150
eyepiece: Huyghenian, 44
eyepiece: Kellner, 44
eyepiece: maximum focal length, 42, 43, 147
eyepiece: micrometer, 42, 153, 169, 170

eyepiece: monocentric, 44
eyepiece: Nagler, 44, 45
eyepiece: orthoscopic, 44
eyepiece: parfocal, 43, 45
eyepiece: Plössl, 44
eyepiece projection, 160, 161
eyepiece: Ramsden, 44
eyepiece: wide angle, 42, 45
eye: pupil, 32
eye relief, 34, 35, 43
eye: resolution, 33, 179

Fabry lens, 214, 215
false observations, 174
fibre optics, 22, 221
field curvature, 20, 49, 51
field of sharp focus, 18, 20
field of view, 16, 35, 42, 150
field of view: eyepiece, 35, 150
figure: of a mirror , 15
figuring: mirrors, 39, 63
filled aperture, 29, 54
filter, 47, 160, 208
filter: comet, 48
filter: H-α , 162
filter: H-α cut-off, 48
filter: light pollution rejection, 48
filter: LPR: see filter: light pollution rejection,
filter: nebula, 48
filters: UBV, 208, 212, 216
finder chart, 149
finder telescope, 148, 149, 158
finding, 42, 92, 132, 148, 150, 158, 172
first point of Aries, 87, 88, 90, 91, 93, 98, 104, 111
five parts rule, 97
fixing a photographic emulsion, 188
Flamsteed, John, 163
Flamsteed system of stellar nomenclature, 163
flat fielding, 185, 194
flexure, 55, 56, 134
fluorescent coating: charge coupled device, 184
focal length, 42, 43
focal ratio, 8, 14
focus correction formula, 174
fog in a photographic emulsion, 209
fork telescope mounting , 126, 128, 129, 135
Foucault test, 39, 40, 63
four parts rule, 97
fovea centralis, 37, 177, 179
Fraunhofer , Joseph von, 14
frequency, 139
full aperture solar filter, 47, 159, 160

galactic latitude, 93
galactic longitude, 93
galaxies: observing, 171
galaxies: visual appearance, 171
Galilean refractor, 3, 4, 5

Galilean satellites, 5
Galileo, 3, 7
German telescope mounting , 126,
127, 135
gibbous phase, 116
glass: crown, 13
glass: flint , 13
glass: optical quality, 7, 14
globular cluster, 169
GMT: see Greenwich mean time,
Gosset, William, 203
great circle, 95
greatest elongation, 116, 117, 155
Greek alphabet, 251
Greenwich / local sidereal times:
relationship, 88
Greenwich mean time, 87
Greenwich meridian, 74, 75, 83, 85
Greenwich sidereal time, 88, 89
Gregorian calendar, 94
Gregorian telescope, 9, 10, 11, 12
Gregory, James, 9
guide star, 24
guide telescope, 134
guider: auto, 134
guiding, 41, 133, 134, 221
guiding: off-axis, 134
gyro-stabilised binoculars, 69

Hadley, John, 12
HA: see hour angle,
Haig mounting, 57
half-wave dipole, 26
Hall, Chester Moor, 12
Hartmann sensor, 24
Harvard system of spectral
classification, 223
Hawaii, 58
heliocentric Julian date, 94
heliocentric theory, 5, 7
heliocentric time, 93
Herschel, John, 165, 173
Herschel wedge: see solar diagonal,
Herschel, William, 12, 15, 38, 165,
173
Herschelian telescope, 12, 17
Hevelius, Johannes, 8
Hipparchus of Nicaea, 110, 165
Hipparcos: spacecraft , 111
history of telescopes, 21
honeycomb mirror, 40
horizon: true, 77, 78
hot spots: charge coupled device
images, 186
hour angle, 56, 82, 83, 84, 85, 88, 91,
102, 126, 132, 133
hour angle / local sidereal time /
right ascension relationship, 91,
92
hour angle / longitude relationship,
85
hour: angular measure , 83
HST: see Hubble space telescope,
Hubble space telescope, 11, 23, 24,
34, 38, 39, 64
Huyghenian eyepiece, 44
Huyghens, Christian, 7, 8
hyperboloidal mirror, 16

H-α cut-off filter, 48
H-α filter, 162

image processing, 184, 185, 186
image processing: contrast
stretching, 185, 195
image processing: dark signal
subtraction, 185, 194
image processing: flat fielding, 185,
194
image processing: hot spots, 186
image processing: sky noise
subtraction, 185
image processing: smoothing, 195
image processing: unsharp
masking, 194
image rotation, 57, 129
inferior conjunction, 116, 117
inner planet, 116, 117
instrumental profile, 195
integrated magnitude, 214
interference fringes, 52
interferometer, 25, 27, 52
interferometer: multiple telescopes,
28, 30
interferometer: optical, 52
interferometer: radio, 52
interferometer: resolution, 53
interferometer: very-long-base-
line, 27, 111
intermediate frequency, 26
international atomic time, 87
invention: telescope, 3
irradiance of a 6th magnitude star,
166
irradiation (see also tear drop
effect) , 177

James Webb telescope, 23
jansky, 25, 140
Jansky, Karl, 25
Josephson junction, 190
Julian calendar, 94
Julian date, 93, 94
Julian date: heliocentric, 94
Julian day number, 94

Keck telescope, 11, 22, 23, 32, 39
Kellner eyepiece, 44
Kepler, Johannes, 110
Kitt Peak, 58

La Silla, 58
latent image in a photographic
emulsion, 188
latest evening, 104
latest morning, 104
latitude, 74, 75, 102
latitude: celestial, 93
latitude: galactic, 93
least squares curve fit: see linear
regression,
lens: achromatic, 12, 13, 14, 49
lens: apochromatic, 13, 14
lens: converging, 13

lens: correcting, 17, 20
lens: diverging, 13
lens: simple, 6, 7, 14
libration, 108, 153
light grasp, 32, 35, 68
light pollution rejection filter, 48
light: velocity of, 139, 140
light: wave nature, 32
limiting magnitude, 168, 171
limits on eclipses, 120
linear regression, 198, 199, 200, 202,
203
Lippersheim, Hans, 3
Lippershey, Hans, 3
local / Greenwich sidereal times:
relationship, 88
local oscillator, 26
local sidereal time, 88, 91, 92, 102
local sidereal time: calculation of,
89
local sidereal time / right ascension
/ hour angle relationship, 91, 92
longitude, 74, 75, 84, 85, 88, 89
longitude: celestial, 93
longitude: galactic, 93
LPR filter: see light pollution
rejection filter ,
luminosity class, 225
luminosity: stellar, 213
lunar eclipse, 119, 121, 152
lunar motion, 107
lunar phase, 108, 116, 118, 151
lunar rotation: tidally-locked, 107,
108

magnetic poles, 77
magnification: minimum, 33, 35,
36, 42, 43
magnification: telescope, 31
magnitude: absolute, 166, 210, 211,
213
magnitude: apparent, 166, 167, 210,
211
magnitude: B, 212
magnitude: bolometric, 213
magnitude: eye estimate, 168
magnitude: integrated, 214
magnitude: limiting, 168
magnitude: photographic, 211
magnitude: photo-visual, 212
magnitude: stellar, 165, 167
magnitude: U, 212
magnitude: V, 212
magnitude: visual, 211, 213
magnitude: wavelength
dependence, 211
making your own telescope
mounting, 129
Maksutov telescope, 20, 21, 67
martian canals, 175, 177
Mauna Kea, 34
maximum entropy methods, 195
maximum focal length: eyepiece,
42, 43, 147
mean solar time, 85
mean sun, 86
mean value, 195, 196
Medicean stars, 5

MEM: see maximum entropy methods,
Mercury: phase, 116, 119
meridian: Greenwich, 74, 75, 83, 85
meridian: prime , 82, 83, 84
MERLIN, 2, 59
Messier objects, 149, 172
metal-on-glass mirror, 16, 38
microdensitometer, 209, 210, 214
micrometer eyepiece, 42, 153, 169, 170
micrometer eyepiece: calibrating, 169
micrometry, 42
minimum magnification, 33, 35, 36, 42, 43
minute: angular measure , 83
mirror blank, 39, 63
mirror cell, 65
mirror: correcting, 24
mirror: figure, 15
mirror: figuring, 39, 63
mirror: honeycomb, 40
mirror: hyperboloidal, 16
mirror: manufacture, 38, 39, 63
mirror: metal-on-glass, 16, 38
mirror: monolithic, 22
mirror: off-axis, 12
mirror: paraboloidal, 9, 10, 11, 12, 17, 19, 63
mirror: polishing, 39, 63
mirror: primary, 9, 10, 12
mirror: secondary, 10, 12, 64
mirror: segmented, 22, 23
mirror: spherical, 6, 19
mirror: stress figuring, 40
mirror: surface accuracy , 38, 64, 157
mirror: thin, 41
mirror: tip-tilt, 24
misalignment of a mounting, 149
MKK system of spectral classification, 223
modified English telescope mounting , 126, 127
modified Julian date, 94
monocentric eyepiece, 44
monolithic mirror, 22
moon, 107, 150
moon maps, 151
Moore, Sir Patrick, 173
motion: annual, 103, 105, 106
motion: direct, 108, 109, 110
motion: retrograde, 108, 109, 110
Mount Hopkins observatory, 22
Mount Palomar, 40, 74
mounting: alt-az, 18, 55, 56, 57, 91, 126, 129
mounting: ball and claw, 135
mounting: barn door, 57
mounting: Dobsonian, 67, 68, 129, 130
mounting: equatorial, 18, 55, 56, 91, 126
mounting: Haig, 57
mounting: scotch, 57
mounting: single pivot, 135
mounting: telescope, 17, 54, 67, 125, 135

movement of the planets, 108, 109, 110
multi-mirror telescope, 20, 21, 22
multi-object spectroscopy, 221

nadir, 77
Nagler eyepiece, 44, 45
Nasmyth telescope system, 18, 19
nebula filter, 48
nebulae: observing, 171
nebulae: visual appearance, 171
negative image: photographic emulsion, 188
New General Catalogue, 172
new moon in the old moon's arms, 152
Newtonian telescope, 10, 11, 15, 16, 62, 64, 66, 67, 68
Newton, Isaac, 10, 38
NGC: see New General Catalogue,
night vision glasses, 69
noise reduction, 195
nomenclature: asteroids, 163
nomenclature: comets, 163
nomenclature: stellar, 162, 164
nomenclature: variable stars, 163
north celestial pole, 75, 76

objective, 5, 8, 13, 15, 34, 38
observations: false, 174
observatory, 58
observing during the day, 173
observing site, 58, 146
observing site: requirements for, 58
observing techniques: photometry, 215, 216
observing: visual, 145
occultation, 98, 99, 117, 120, 154
off-axis guiding, 134
off-axis mirror, 12
operating principle of a charge coupled device, 180, 181
operating principle of a p-i-n photodiode, 189
opposition, 116, 117
optical aberration , 5, 13, 14, 19, 48
optical density of a photographic emulsion, 210
optical interferometer, 52
optics: telescope , 31
orbital period: sidereal, 121, 122
orthoscopic eyepiece, 44
outer planet, 116, 117
over-coating: silicon dioxide, 16, 64

paraboloidal mirror, 9, 10, 11, 12, 17, 19, 63
parallax, 113, 115, 116
parfocal eyepieces, 43, 45
Parsons, William, 15
partial solar eclipse, 118, 120
periastron, 121
pericynthion, 121
perigee, 121
perihelion, 121
perijove, 121

phase: moon, 108, 116, 118, 151
phases of Mercury, 116, 119
phases of Venus, 5, 116, 119
photographic emulsion, 187
photographic emulsion: characteristic curve, 195, 209, 210
photographic emulsion: detection principle, 187
photographic emulsion: developing, 188
photographic emulsion: electron trap, 187
photographic emulsion: fixing, 188
photographic emulsion: fog, 209
photographic emulsion: latent image, 188
photographic emulsion: negative image, 188
photographic emulsion: optical density, 210
photographic emulsion: prints, 189
photographic emulsion: processing, 188
photographic emulsion: resolution, 187
photographic emulsion: spectral sensitivity, 187
photographic emulsion: speed, 187
photographic magnitude, 211
photographic photometry, 208, 209, 210
photography, 180, 186
photography: colour images, 189
photometer, 189, 214
photometric system: UBV, 212, 213, 219
photometry, 144, 207
photometry: charge coupled device, 207, 208
photometry: observing techniques, 215, 216
photometry: photographic, 208, 209, 210
photomultiplier, 214
photon, 141
photo-visual magnitude, 212
pin-cushion distortion, 51, 52
p-i-n photodiode, 180, 189, 207, 214
p-i-n photodiode: operating principle, 189
planet: inner, 116, 117
planet: outer, 116, 117
planetary positions: relative, 116
planets: best telescope for observing, 156
planets: movement of, 108, 109, 110
planets: observing, 154, 155, 156
planisphere, 148
Plössl eyepiece, 44
Pogson's equation, 165
Pogson, William, 165
point source, 31
point spread function, 195
polar axis, 56, 126
polarimetry, 144
polarisation of sky light, 173
polarised electromagnetic radiation, 141

polarised electromagnetic
    radiation: circular, 141
polarised electromagnetic
    radiation: elliptical, 141
polarised electromagnetic
    radiation: plane, 141
pole star, 112
polishing: mirrors, 39, 63
Porta, Giambattista della, 5
portable telescope, 131
position angle, 169, 170
position in an orbit, 121
position in the sky, 73
precession, 111, 112, 113, 114, 149
primary mirror, 9, 10, 12
prime focus, 17, 25
prime meridian, 82, 83, 84
printing a photographic negative,
    189
prism, 220, 221
processing a photographic
    emulsion, 188
prominence spectroscope , 163
proper motion, 111
Ptolemy, 109, 110
pupil of the eye, 32
purchasing a telescope, 67
Pyrex, 40, 63

quadrature, 117, 118
quantum: of radiation , 141

radial velocity, 225
radio interferometer, 52
radio receiver: super-heterodyne,
    26
radio telescope, 25, 26
Ramsden eyepiece, 44
RA: see right ascension,
rear illuminated charge coupled
    device, 184
reflecting telescope , 9, 10
reflective coating: aluminium, 16,
    64
reflective coating: silver, 16, 64
refraction, 13
refraction: atmosphere, 105, 134
refractive index, 139
refractor: astronomical, 5, 15
refractor: Galilean, 5
refractor: Yerkes, 14, 15
relative planetary positions, 116
relay lens, 8
resolution, 23, 24, 25, 27, 32, 33, 53,
    169, 179, 187, 219, 220
resolution: atmospheric limitation,
    34, 147
resolution: eye, 179
resolution: interferometer, 53
resolution: photographic emulsion,
    187
resolution: spectral, 219, 220
retina, 31, 37, 177, 179
retrograde motion, 108, 109, 110
rhodopsin, 37, 178
right ascension, 56, 90, 91, 102, 126,
    132, 133

right ascension / local sidereal time
    / hour angle relationship, 91, 92
rising azimuth, 102
rising time, 102
Ritchey-Chrétien telescope, 16, 18,
    68
rod cells: retina , 31, 37, 177, 179
Rómer, Ole, 120

safety: while observing , 146, 158,
    160, 174
Saturn's rings, 7, 9
Schmidt, Bernhard, 18
Schmidt camera, 18, 19, 20, 38
Schmidt-Cassegrain telescope, 20,
    21, 61, 62, 67
Schottky diode, 26
scintillation, 23, 147
scotch mounting, 57
seasons, 103, 105, 106
second: angular measure , 83
secondary mirror, 10, 12, 64
secondary mirror: ellipsoidal, 9
secondary mirror: flat, 10, 64, 65
secondary mirror: hyperboloidal,
    11, 16
second hand telescopes, 62
Seebeck effect, 189
seeing, 147
segmented mirror, 22, 23
Serrurier truss, 55
setting azimuth, 102
setting circle, 91, 92, 126, 132, 133,
    147, 148, 155, 158, 172
setting time, 102
setting (a telescope on an object) ,
    42, 92, 132, 147, 148, 150, 158, 172
shadow circularisation: finding the
    sun, 158, 159
Short, James, 12
SI units, 249
sidereal / synodic periods:
    relationship, 122
sidereal day, 56, 82, 89, 91, 101, 126
sidereal orbital period, 121, 122
sidereal time, 87, 88, 89, 148
sidereal time: Greenwich, 88, 89
silicon dioxide over-coating, 16, 64
silver bromide , 187
silver mirror test, 64
sine rule: Euclidean , 95
sine rule: spherical , 97
single pivot telescope mounting,
    135
sky light: polarisation of, 173
sky noise subtraction, 185
smoothing, 195
solar day, 82, 89
solar diagonal, 45, 46, 47
solar eclipse, 118, 120
solar filter: full aperture, 47, 159,
    160
solar telescope, 161
solar time, 86
solar time: mean, 85
solstice, 82, 104
south celestial pole, 75, 76
space telescope, 20, 21, 23

spectral class , 222, 223, 224, 225
spectral class: Harvard system, 223
spectral class: MKK system, 223
spectral resolution, 219, 220
spectral response of a charge
    coupled device, 184
spectral sensitivity of a
    photographic emulsion, 187
spectral type: see spectral class,
spectroscope: design, 220
spectroscope: direct vision, 222
spectroscopy, 144, 219, 222
spectroscopy: multi-object, 221
speculum metal, 11, 15, 16
speed of a photographic emulsion,
    187
spherical aberration, 6, 7, 8, 9, 12,
    19, 48, 49
spherical mirror, 6, 19
spherical polar coordinates, 73, 74,
    75, 80, 91
spherical triangle, 97
spherical trigonometry, 95, 97
spider, 65
spring equinox, 82, 104
standard deviation, 195, 196
standard error of the mean, 195,
    196, 197
standard star, 166
star: binary, 169
star: colour, 168
star: comparison, 215
star diagonal, 45
star: distance, 114
star: double, 168, 169, 170
star hopping, 150, 172
star: observing, 162, 166
star: standard, 166
star: visual binary, 170
statistics, 198, 199, 200, 201, 202,
    203, 204
stellar aberration, 114, 115, 116
stellar luminosity, 213
stellar magnitude, 165, 167
stellar nomenclature, 162, 164
stellar nomenclature: Bayer system,
    163, 164
stellar nomenclature: Flamsteed
    system, 163
STJs: see superconducting tunnel
    junction detector,
stopping down a telescope, 160
stress figuring: mirrors, 40
Student's t test, 198, 203, 204, 205
summer solstice, 82, 104
summer triangle, 104
sun: finding, 158
sun: observing, 158
sun: warning regarding observing,
    158, 174
sunspot number: see Zurich
    sunspot number
superconducting tunnel junction
    detector, 26, 190, 219
superconductor, 190
super-heterodyne radio receiver,
    26
superior conjunction, 117
support: active, 55

surface accuracy of mirror , 38, 64, 157
surface brightness of extended objects, 36, 37
synodic period, 121, 122
synodic / sidereal periods: relationship, 122

TAI: see international atomic time,
tear-drop effect, 175, 177, 179
telecompressor, 47
telephoto property, 11, 19, 20
telescope, 3, 31, 61
telescope: aerial, 8
telescope: Cassegrain, 11, 18, 19, 22, 25, 68
telescope: catadioptric, 19
telescope: catoptric, 19
telescope: Digges', 4
telescope: dioptric, 19
telescope: Dobsonian, 67, 68
telescope: finder, 148, 149, 158
telescope: Gregorian, 9, 10, 11, 12
telescope: guide, 134
telescope: Herschelian, 12, 17
telescope: history, 21
telescope: invention , 3
telescope: magnification, 31
telescope making, 62, 63
telescope: Maksutov, 20, 21, 67
telescope mounting, 17, 54, 67, 125, 135
telescope mounting: active, 54, 55
telescope mounting: English, 126, 128
telescope mounting: fork, 126, 128, 129, 135
telescope mounting: German, 126, 127, 135
telescope mounting: making your own, 129
telescope mounting: modified English, 126, 127
telescope mounting: requirements, 125
telescope mounting: yoke, 126, 128
telescope: multi-mirror, 20, 21, 22
telescope: Newtonian, 10, 11, 15, 16, 62, 64, 66, 67, 68

telescope optics, 31
telescope: portable, 131
telescope: purchasing, 67
telescope: radio, 25, 26
telescope: reflecting, 9, 10
telescope: Ritchey-Chrétien, 16, 18, 68
telescope: Schmidt-Cassegrain, 20, 21, 61, 62, 67
telescope: second hand, 62
telescope: solar, 161
telescope: space, 20, 21, 23
telescope suppliers, 229
telescope system: Coudé, 18
telescope system: Nasmyth, 18, 19
telescope: terrestrial, 8
telescope tube, 54, 55, 66, 125
terrestrial telescope, 8
thin mirror, 41
thinned charge coupled device, 184
tidally-locked lunar rotation, 107, 108
tide, 108
time, 85
time: Greenwich mean, 87
time: heliocentric, 93
time: international atomic, 87
time: local sidereal, 88, 91, 92, 102
time: sidereal, 87, 88, 89, 148
time: universal, 87
tip-tilt mirror, 24
TLP: see transient lunar phenomenon,
tool, 39, 63
total solar eclipse, 118, 120
track, 56, 125
transient lunar phenomenon, 153, 154
transit, 117, 119, 175
triangle: spherical, 97
true horizon, 77, 78
twilight, 104, 105
twinkling, 147, 156

UBV filters, 208, 212, 216
UBV photometric system, 212, 213, 219
ULE, 40, 63
U magnitude, 212

uncertainties in data, 195, 196, 197
uncertainties: estimating, 197
unfilled aperture, 29, 30
universal time, 87
unsharp masking, 194
UT: see universal time,

variable star nomenclature, 163
velocity of light, 139, 140
Venus: phases, 5, 116, 119
vernal equinox, 82, 87, 104
very large array, 29
very-long-base-line interferometry, 27, 111
visual appearance of galaxies, 171
visual appearance of nebulae, 171
visual binary star, 170
visual magnitude, 211, 213
visual observing, 145
visual purple, 37, 178
VLA: see very large array,
VLBI: see very-long-base-line interferometry,
V magnitude, 212

wavelength, 139
wavelength dependence of stellar magnitudes, 211
wide angle eyepiece, 42, 45
winter solstice, 82, 104

Yagi antenna, 26
Yerkes refractor, 14, 15
yoke telescope mounting , 126, 128

zenith, 77, 82, 105
zenith distance, 80
Zerodur, 40, 63
zodiac, 78, 79
zodiacal constellations, 79
zodiacal constellations: dates of solar passage, 79
Zurich sunspot number, 161